The Global Old Age Care Industry

Vincent Horn
Cornelia Schweppe
Anita Böcker • María Bruquetas-Callejo
Editors

The Global Old Age Care Industry

Tapping into migrants for tackling the old age care crisis

palgrave
macmillan

Editors
Vincent Horn
Institute of Education
Johannes Gutenberg University of Mainz
Mainz, Germany

Cornelia Schweppe
Institute of Education
Johannes Gutenberg University of Mainz
Mainz, Germany

Anita Böcker
Faculty of Law
Radboud University Nijmegen
Nijmegen, The Netherlands

María Bruquetas-Callejo
Center for Migration Law/Sociology of Law
Radboud University Nijmegen
Nijmegen, The Netherlands

ISBN 978-981-16-2236-6 ISBN 978-981-16-2237-3 (eBook)
https://doi.org/10.1007/978-981-16-2237-3

This Palgrave Macmillan imprint is published by the registered company Springer Nature Singapore Pte Ltd.
The registered company address is: 152 Beach Road, #21-01/04 Gateway East, Singapore 189721, Singapore

Acknowledgments

This publication is based on the International Conference "The Long-Term Care Crisis: Tapping into Labour Resources Within and Across National Borders" which took place at the University of Mainz (Germany), Institute of Education, March 11–12, 2019. The Conference was financed by the German Research Foundation (DFG). We would like to thank the DFG for its generous support.

Contents

Notes on Contributors

Gudrun Bauer is a social policy expert at the City of Vienna, Austria. In her Ph.D. and while previously working as a research associate in the Department of Socioeconomics at WU Vienna University of Economics and Business, she focused on transnational migrant care work arrangements.

Marianne van Bochove is a sociologist and Assistant Professor of Health Care Governance at Erasmus School of Health Policy and Management, Erasmus University Rotterdam, the Netherlands. Her research and teaching interests and experience include health care policy reform, voluntary work in health and social care, migrant care work and urban citizenship. Across these fields, she is particularly interested in how boundaries between different groups are constructed and how identities are formed and changed.

Anita Böcker is Associate Professor of Sociology of Law in the Faculty of Law at Radboud University, Nijmegen, the Netherlands. She has written widely on the social and legal status of immigrants, the effectiveness of anti-discrimination law and the regulation of migration and immigrant integration in the Netherlands and other European states. Her recent research focuses on migration and return migration in later life and on live-in migrant care arrangements in Germany and the Netherlands.

María Bruquetas-Callejo is a research fellow at the Center for Migration Law/Sociology of Law, Radboud University, Nijmegen, the Netherlands. Her work deals with migration and welfare state policies from a comparative, multidisciplinary and multi-level perspective. She has written widely on immigrants' access to health care and to the welfare state, on immigrants' integration policies and on transnational eldercare arrangements. Her recent research examines the migration-welfare state nexus, emphasizing the contribution of migrants to retrenching welfare states in the field of long-term care.

Chiara Giordano is working as a postdoctoral researcher at the Group for Research on Ethnic Relations, Migration and Equality (GERME), University of Brussels (ULB), Belgium. She holds a Ph.D. in Economic Sociology and Labor Studies/*doctorat en sciences sociales et politiques*, which she has accomplished in co-tutorship between the University of Milan (Università Statale di Milano) and the University of Brussels (Université Libre de Bruxelles). She also holds a Master's degree in Gender Studies and Equal Opportunities. Her research interests include female migration, domestic and care work, gender inequality and ethnic discrimination.

Anita Ham is a Ph.D. candidate in the Department of Public Health and Primary Care at Leiden University Medical Center, the Netherlands. Her research is affiliated with the Anthropology and Sociology Department of University of Amsterdam. Her research, funded by the Dutch Organization for Scientific Research (NWO) (023.008.044/1105), focuses on diversity and inclusivity, integration and aging citizens. She is a registered nurse, lecturer and researcher at The Hague University of Applied Sciences, Faculty of Health, Nutrition and Sport (Department Nursing). She received her M.Sc. in Medical Anthropology and Sociology from the University of Amsterdam in 2009.

Vincent Horn is a postdoctoral researcher at the Institute of Education, Johannes Gutenberg University of Mainz, Germany. His expertise includes transnational aging, transnational migration in old age, old age and transnational families, old age care migration and migrant care migration as well as older refugees. He is a postdoctoral researcher in the

Project "Emergence and Significance of Transnational Old Age Care Arrangements in Germany and the Netherlands", financed by the German Research Foundation (DFG). In addition, he is conducting a study on older refugees in Germany.

Duco zur Kleinsmiede is a control specialist at one of the biggest health insurers in the Netherlands. He holds Master's degrees in Communication and Information Science from Tilburg University, and Healthcare Management from Erasmus University Rotterdam (EUR). His Master thesis at EUR was about the boundary crossings of live-in migrant care workers in the Netherlands. He explored how care workers combine their lives 'here', in the Netherlands, and 'there', in the country of origin, and how they construct borders between their 'front stage' work and 'back stage' private lives in the client's home.

Simone Leiber is Professor of Social Policy at the University of Duisburg-Essen, Germany. Her research interests are comparative welfare state reform, long-term care policies as well as the interrelation of social policy and social work.

Li-Fang Liang is an assistant professor in the Department of Sociology at National Dong Hwa University, Taipei, Taiwan. She received her Doctoral degree from the Department of Sociology at Syracuse University. She has conducted research on how the social organization and social relations of care work (re)shape the lived experiences of live-in migrant care workers and care recipients. Her research explores how migrant care workers maintain their care responsibilities and practices in the transnational context.

Claudia Menebröcker studied sociology and educational science. She currently works as a consultant for elderly care for the Caritas Association for the Archdiocese of Paderborn, Germany. She is the contact person for the coordinators of the CariFair project, with which Caritas is committed to fair and legal employment of live-ins. Together with the coordinators she works on good working conditions and suitable qualifications of the carers who are employed in German families within the framework of CariFair.

Reiko Ogawa is a professor at the Graduate School of Social Sciences, Chiba University, Japan. Her research focuses on migration and care work. Some of her recent publications include "Making of Migrant Care Workers in East Asia", in *Routledge Handbook on Gender in East Asia* (2020); *Gender, Care and Migration in East Asia* (co-edited, Palgrave Macmillan, 2018); and "Japan: From Social Reproduction to Gender Equality", in *Women, Work and Care in the Asia-Pacific* (2017).

Antero Olakivi is a postdoctoral researcher in the Centre of Excellence in Research on Ageing and Care (funded by the Academy of Finland, #312310) at the University of Helsinki, Finland. His sociological Ph.D. dissertation, "The Relational Construction of Occupational Agency: Performing Professional and Enterprising Selves in Diversifying Care Work" (University of Helsinki, 2018), examined the construction (and silencing) of ethnic and migration-based inequalities in eldercare work in Finland. He has written articles in journals such as *Sociology of Health and Illness* and *Journal of Professions and Organization*.

August Österle is an associate professor in the Department of Socioeconomics at WU Vienna University of Economics and Business, Austria, and a visiting professor at Corvinus University of Budapest, Hungary. His research and teaching focuses on comparative welfare state research, in particular in the field of long-term care, on transnational social policy and on access to social rights.

Verena Rossow is a research associate at the Institute for Economics, Labour and Culture (IWAK), Goethe University Frankfurt/M, Germany. Her research foci are on domestic care, marketization and commodification of intimate workspaces, just as live-in care work.

Cornelia Schweppe is Professor of Social Pedagogy at the Johannes Gutenberg University of Mainz, Germany. She was the director of the Doctoral Research Program "Transnational Social Support" financed by the German Research Foundation (2008–2017) and was awarded the five-year fellowship at the Gutenberg Research College (2014–2019). She has written widely on transnational aging, retirement migration and old age care. Her research includes projects on retirement migration to Kenya and Thailand, care migration in old age to Poland and Thailand and transnational care arrangements in Germany and the Netherlands.

Bernhard Weicht works in the Department of Sociology at the University of Innsbruck, Austria. He studied Economics in Vienna and Social Policy in Nottingham. He holds a Ph.D. from the University of Nottingham where he researched the social and moral construction of care. He continued his work at Utrecht University with a project on the intersections of care and migration regimes. He received his *Habilitation* in 2018. Weicht has written on the construction of care, ideas of dependency, migrant care workers, migration and care regimes and the construction of aging. He is the author of *The Meaning of Care* (2015).

Sirpa Wrede is Professor of Sociology at the University of Helsinki, Finland. Her research interests deal with the dynamics of inequality in the social organization of social service professionalism and social citizenship. Her work focuses on aging and social ties in the context of migration and on professional groups in the context of neoliberal globalization. She leads a research team within the Centre of Excellence in Research on Ageing and Care (funded by the Academy of Finland, #312310), of which she is the Vice Director.

List of Figures

1

Introduction: The Global Old-Age Care Industry—Tapping into Care Labor Across and Within National Borders

Vincent Horn, Cornelia Schweppe, Anita Böcker, and María Bruquetas-Callejo

A few weeks after the first COVID-19 infections were diagnosed in Germany, NGOs and experts in the field of old-age care warned of acute bottlenecks in the provision of care for older Germans. They feared that care workers from Eastern Europe who were travelling back to their home countries might not be able to return to Germany due to the closure of borders. As a result, up to 200,000 migrant care workers could be

V. Horn (✉) • C. Schweppe
Institute of Education, Johannes Gutenberg University of Mainz, Mainz, Germany
e-mail: hornv@uni-mainz.de; c.schweppe@uni-mainz.de

A. Böcker
Faculty of Law, Radboud University Nijmegen, Nijmegen, The Netherlands
e-mail: a.bocker@jur.ru.nl

M. Bruquetas-Callejo
Center for Migration Law/Sociology of Law, Radboud University Nijmegen, Nijmegen, The Netherlands
e-mail: m.bruquetascallejo@jur.ru.nl

V. Horn et al. (eds.), *The Global Old Age Care Industry*,
https://doi.org/10.1007/978-981-16-2237-3_1

missing after Easter.[1] The weekly German magazine Der Spiegel quoted one NGO's spokesperson saying: "We already had a shortage of care workers, the crisis has been here for a long time." Another NGO's spokesperson was particularly concerned about older people in need of care who were living at home with a live-in migrant carer: "Where should all these people stay? Old-age and nursing homes are full." This NGO asked for financial compensation for people in paid work who reduced their working hours or temporarily left their job to care for relatives who were left without care due to the coronavirus pandemic (Der Spiegel, March 24, 2020; Lehmann, April 3, 2020). Such a compensation scheme did not materialize, but Germany and several other EU countries have temporarily relaxed their national qualification requirements for care professions in order to tap into another source of care labor: refugees and other recently arrived migrants (Wilfried Martens Centre for European Studies, May 29, 2020).

This episode from Germany illustrates the worries and consequences related to the sudden immobility of migrant care workers from neighboring countries during the COVID-19 pandemic. Although the topic soon disappeared from the public discourse, the episode emphasizes the heavy reliance of the country's old-age care system on care labor from abroad. Especially private households depend on the support of circular migrants working as live-in carers for older people (Horn et al., 2019). These live-in carers are usually not trained nurses or otherwise certified as carers for older people. Moreover, most of them are employed irregularly by German families. Although the legal framing differs from one country to another (Steiner et al., 2019), transnational live-in migrant carer arrangements have also become established in Austria and Switzerland (Chau, 2020; Österle & Bauer, 2016). In all three countries, the COVID-19 pandemic led to very similar concerns among caring relatives and civil society organizations. Analyzing policy responses to impending shortages in old-age care, Leiblfinger et al. (2020: 147) conclude that "various measures were implemented to ensure that Austrian, German, and Swiss elder population did not have to do without their live-in carers". These measures, however, often went at the expense of the migrant care workers who either had to prolong their stay, risk an infection on their trips back (-and-forth) or quit their job "at the detriment of their economic survival and that of their families" (Giordano, 2020: 1).

The episode also indicates that labor shortages in Germany's old-age care have already been critical prior to the COVID-19 pandemic. In fact, many high-income countries experience labor shortages in their old-age care systems for a long time. To tackle these shortages, looking for both trained and untrained care workers abroad has become a widespread strategy of governments, private care providers and brokerage agencies in many countries, such as in Australia (Howe, 2009; Morrison-Dayan, 2019), Italy (Ambrosini, 2013; Degiuli, 2007), Germany (Leiber et al., 2019), Spain (León, 2010) and the United Kingdom (Cangiano & Walsh, 2014; Anderson & Shutes, 2014), as well as in countries in East Asia (Asis & Piper, 2008; Huang et al., 2012; Michel & Peng, 2017; Raghuram, 2012; Yeoh & Huang, 2010) and North and South America (Chaney, 1991; Fudge & Parrott, 2014; Hondagneu-Sotelo, 2001; Lobel, 2001; Parreñas, 2001). Simultaneously, the political and private sector are doing increasing efforts to recruit and/or train migrant and refugee populations, both settled and recently arrived, for oftentimes low paid and undervalued jobs in old-age care. In this book, we therefore distinguish between tapping into care labor across and within national borders in order to fully grasp the role of migration for old-age care systems around the globe.

Focusing the processes of tapping into care labor across and within national borders, this book aims at examining the increasing complexity and entanglements of migration and old-age care at the global level. It explores the mechanisms, policies, strategies and actors that are involved in mobilizing, training, recruiting and placing migrant care workers from economically poorer countries in private households and old-age care services in economically richer countries. It looks at both state-driven recruitment programs and private-sector initiatives, such as for-profit brokerage and employment agencies. Likewise, other actors such as families, public care service providers and welfare associations are considered, along with their increasing influence concerning the way and extent to which migrants are channeled into specific types and forms of employment in an emerging global old-age care *industry*. In addition, more and more diverse and complex patterns of old-age care delivery are surfacing when care labor is tapped from abroad. Therefore, this publication also addresses questions regarding how old-age care is negotiated, what old-age care concepts emerge and what implications they might have for the well-being and social protection of the different actors involved. Thus, it delves into the analysis of different levels of this phenomenon, ranging from policies and strategies to promoters and intermediaries to care constellations and relationships and implications for the well-being and social protection of the different actors involved. These processes will be highlighted from different countries and regions of the world.

In fact, a myriad of ways exists in how the relationships between migration and old-age care are framed institutionally between countries and regions. The degree of formalization, for example, can vary considerably from one country to another. While in some the recruitment and employment of migrants for old-age care are highly regulated by migration and labor law, such regulations can vastly be absent in others. Moreover, a co-existence of formal and informal systems of old-age care involving migrants can be observed in one and the same country (see Österle and Bauer in this book). Related to this, there is a large variety of regulations regarding the professional skills of migrants working in old-age care. Working in formal old-age care settings generally requires professional skills, obtained in the sending and/or receiving country. By contrast, working as a live-in carer in the shadow economy does not require training as a nurse or the like. All of this has implications for both the quality and safety of care delivered and the interactions between migrant care workers, dependent older people and family caregivers.

In the following, we will first outline how cross-country differences regarding the role of migration for old-age care can be theoretically underpinned by drawing on literature at the intersection of care, migration and employment regimes. Second, we set the scene for this book by proposing the perspective of the global care *industry* as an analytical tool for studying the multiple levels and actors involved in the configuration and perpetuation of migration-based old-age care arrangements.

1 The Migration-Care Regime Nexus

Research has shown that the scope and incorporation of migrant old-age care workers differ enormously among countries. These differences have been explained by variations in old-age care regimes, as well as historical links, geographic proximity and patterns and regimes of migration (Anderson, 2012; Da Roit & Weicht, 2013; Simonazzi, 2009; Theobald, 2011). In the tradition of comparative welfare regime literature (Esping-Andersen, 1999; Lewis, 1992), researchers seek to identify different types of intersecting care, employment and migration regimes, using indicators such as the type and extent of public provisions for old-age care, social insurance schemes, migration and labor market policies, social norms regarding caregiving and so on.

For European countries, van Hooren (2012) differentiates among social-democratic, familistic and liberal old-age care models. This differentiation allows researchers to loosely cluster European countries according to the roles and positions of migrant care workers in their old-age care systems.

In the *social-democratic* model, the state assumes major responsibility in the form of high public expenditure and the provision of a relatively generous public old-age care infrastructure. Private expenditures tend to be low and cash benefits, if available, are tightly regulated. Despite the great availability and affordability of old-age care services, the family remains crucial to old-age care in these countries, although less in the form of hands-on care than in that of practical and emotional support. Compared to the state and the family, private for-profit providers continue to play a minor, though increasingly important, role in the provision of old-age care services (Anderson & Kvist, 2015; Mot et al., 2010; Szebehely & Meager, 2018). Migrant care workers are primarily employed in public or private non-profit care services and, to a lesser extent, by private for-profit providers. These workers, however, are more likely than their native counterparts to find themselves in precarious employment conditions (Behtoui et al., 2020; Isaksen, 2010; Jönson & Giertz, 2013). In contrast, the recruitment of care workers abroad remains very limited and, if taking place, is carried out by private brokerage agencies operating in the field of home-based care (see Horn et al., 2019 for the Netherlands).

As indicated by its name, the family is the central provider of old-age care in the *familistic* model. Compared to the social-democratic model, the public old-age care infrastructure in countries with a familistic old-age care model is less developed and private out-of-pocket costs are relatively high. Old-age care services are mainly offered by third-party and for-profit providers, offering families a limited (because often costly) choice to share their caring responsibility. Cash-for-care benefits exist to varying degrees and their use tends to be uncontrolled. This lack of control has been shown to facilitate the employment of oftentimes unskilled live-in migrant carers (LIMCs) in these countries (Da Roit & Weicht, 2013; Lamura et al., 2010). In fact, European countries with a familistic old-age care model, such as Italy, Spain and Germany, rely heavily upon migrants caring for older people in private households (Ambrosini, 2013; Degiuli, 2007; Horn et al., 2019; Ibáñez & León, 2014). Despite this similarity, the scope and mode of employment (regular vs. irregular) differ considerably across these countries, as do the origins of the LIMCs (EU vs. non-EU countries) and their migration patterns (e.g., permanent vs. pendular migration). In parallel, several of these countries are running programs for recruiting care workers abroad in order to fill labor shortages in the private and public old-age care sectors (e.g., Germany's "Triple Win program").

Finally, in the *liberal* model, the market is an important provider of old-age care, as most services are outsourced to private non-profit and for-profit providers (Van Hooren, 2012). Accordingly, in these countries, families must pay for old-age care services according to their income and assets. Due to means-testing, only less-affluent families are eligible for cash-for-care benefits. As the use of these benefits is controlled, spending this money for an irregularly employed LIMC is not possible (Da Roit & Weicht, 2013). Thus, in the case of the United Kingdom, migrant care workers can mainly be found as formally employed in private-sector residential and homecare services (Anderson & Shutes, 2014; Cangiano & Shutes, 2010; Franklin & Brancati, 2015). Recently, however, live-in care arrangements are becoming more popular in the United Kingdom, with providers praising this arrangement as a suitable solution in the time of COVID-19:

> With so much uncertainty currently surrounding the outbreak of Coronavirus, it's important to do all that we can to keep our elderly

> loved ones safe and mentally well. Live-in elderly care enables older people to continue living independently in their own homes, with the support of a professional live-in carer. With many care homes facing the possibility of being required to lock down, live-in care provides a less disruptive option that allows you to continue visiting loved ones, while also providing peace of mind that they are being looked after should you begin to show signs of Coronavirus yourself. (Independent People Homecare Services, 2020)

Although live-in care arrangements in the United Kingdom are (still) not widely used, it is likely that migrant workers will make up a substantial part of the workforce in this field of the country's old-age care system.

Additionally, in North American countries such as the United States and Canada, migrant care workers play a pivotal role in old-age care (Antanackovic & Bourgeault, 2013; Leutz, 2007; Spencer et al., 2010). When analyzing representative data for the United States, Zallman et al. (2019) found that one out of four workers in the country's formal and non-formal long-term care sector had been born abroad. They conclude that migrants "provide care to hundreds of thousands of people living at home or in other nonmedical settings such as senior housing" and warn that "the current policies that curtail immigration are likely to compromise the availability of care" (924). The United States has increasingly diminished unauthorized immigration and hence the pool of potential care workers, with its visa program offering entry for only a very limited number of (skilled) migrant care workers (Stone, 2016). By contrast, Canada has more strategically used its immigration policies to manage the demand for old-age care workers. The most prominent example is a program for live-in caregivers and its successor program, which offers migrants the possibility to apply for permanent residence when meeting certain requirements (Boyd, 2017). Another example of Canada 'rewarding' care work is their immigration minister Marco Mendicino's recent announcement that Canada will provide asylum seekers who worked in the country's health care sector, and especially in long-term care centers, during the COVID-19 pandemic with a pathway to permanent residence (CIC NEWS, 2020).

Compared to European and North American countries, most East Asian countries experienced demographic challenges and social changes later, but in a more drastic manner (Soma et al., 2011). In fact, the population is aging fastest in this region, as reflected by projections of the increase in the proportion of people aged 65 years and older between 2019 and 2050. Indeed, South Korea (23%), Taiwan (19.9%), Macau (17.7%) and Hong Kong (17.2%) are all amongst the top ten countries expected to experience the largest increase in older people (UN, 2019). As within relatively affluent aging societies, East Asian countries—to varying degrees—are recruiting migrant and foreign-born care workers, predominantly those from Southeast Asian countries. Welfare regimes in East Asia have undergone considerable transformations in the last decades, however, leading to the evolution of different care regimes (Chan et al., 2011; Choi, 2012; Kim, 2015; Michel & Peng, 2017) and different patterns of foreign care worker policies (Hong & Han, 2017; Song, 2015; Wang & Chan, 2017).

Comparing care regimes and migrant care policies in Japan, South Korea and Taiwan, Wang and Chan (2017) reveal differences, but also some similarities, among the three countries. According to these authors, Taiwan was the first country to establish entry routes for mainly low-skilled foreign care workers from specific Southeast Asian countries (e.g., Indonesia, the Philippines and Vietnam), allowing a maximum stay of up to 14 years (thus they are called foreign, and not migrant, care workers). More than 90% of the approximately 240,000 foreign care workers in the country are hired as LIMCs by private households. They do not have permission to choose or transfer between employers and are obliged to pay a monthly 'service fee' to Taiwan-based brokering agencies. Although two 10-year plans for reforming the country's long-term care system have been developed since 2007, thus far a change from dominant LIMC to a social care service model has failed. Chien (2018) explains this failure based on the particular resistance of Taiwanese families with dependent older relatives who fear insufficient social care service development and therefore adhere to the LIMC model.

In contrast to Taiwan, South Korea restricted the entry of migrant care workers to co-ethnic workers from China, 'honoring' its mono-ethnicity

(Wang & Chan, 2017). Most of the between 54,000 and 70,000 foreign care workers work as LIMCs in private households. They are granted five-year multi-entry visas and under certain conditions can settle permanently in the country. After establishing a mandatory long-term care insurance in 2008, the South Korean government aimed to recruit (female) native-born workers to work in the developing public old-age care sector. Nevertheless, "work intensity, low wages and deteriorating working conditions" (Song, 2015: 387) kept native-born workers absent from these jobs. Consequently, uncertified co-ethnic care workers were granted the right to participate in the national qualification test for long-term care workers (Song, 2015). Most co-ethnic care workers, however, remain as LIMCs in the informal niche market.

Of the three countries, Japan has the longest tradition in policies targeting its aging population, beginning in the 1960s (Olivares-Tirado & Tamiya, 2014). The latest important reform took place in 2000, when a universal long-term care insurance was introduced that allowed eligible older persons to choose from a range of long-term care services at home or in institutions (Iwagami & Tamiya, 2019). Benefits for service use are generous, covering 90% of need (Campbell & Ikegami, 2000). Concerns about old-age care in Japan are therefore less related to the availability of social care services and more related to the high costs of its old-age care system (Akemura & Kojima, 2018). Similar to South Korea, the idea of ethnic and cultural homogeneity guides Japans' migration policy. Increasing labor shortages, however, have caused policymakers to slightly open the doors for foreign care workers. Between 2008 and 2012, Japan signed agreements to recruit care workers from Indonesia, the Philippines and Vietnam under the Economic Partnership Agreement (Wang & Chan, 2017). Foreign care workers in Japan can stay up to four years, but may apply for permanent residence and family reunification after passing the national qualification test. In contrast to South Korea and Taiwan, foreign care workers in Japan are only eligible to work in hospitals and nursing homes, not as care providers in private households (see Ogawa in this book). Regardless, while migration is hotly debated, the number of foreign-born care workers in Japan remains very low, accounting for only 2800 people in 2017 (Wang & Chan, 2017). In view of this, Song (2015) doubts if these workers can solve the serious labor shortage in Japan's long-term care sector and predicts no drastic changes in the current model of long-term care provision.

Nevertheless, the question of old-age care provision is not only salient in the above-mentioned countries and regions, but also increasingly in African, Latin American and Caribbean countries as well as in countries in the Middle East (e.g., Israel or Turkey). Formal long-term care systems remain inexistent in many of these countries, but are emerging in some (Acosta González et al., 2018; Lloyd-Sherlock, 2018; Matus-López & Pedraza, 2016; WHO, 2017). Future research will show how old-age care and migration will intersect in these long-term care systems. Will migrant and foreign-born workers, for example, be incorporated into formal long-term care services and institutions, or will they enter and/or remain in the large informal domestic work sector characterizing these regions?

Even though the concept of old-age care regimes and their intersection with migration and employment regimes provides a valuable approach to look at cross-country differences regarding the incorporation of migrant old-age care workers, the concept of regime is highly abstract and difficult to operationalize (Da Roit & Weicht, 2013). The major gap that flaws this literature, however, is the fact that most studies mainly focus on the macro perspective of the institutional regimes, disregarding their relation with the meso level of intermediary organizations and the micro level of migrants' and care recipients' practices. Indeed, the motivations, entry paths and trajectories of migrant old-age care workers are highly diverse and their relationships with the various regimes might be more comprehensively captured by reconstructing their journeys from using a 'bottom-up' approach (for this argument see Weicht in this book).

At the same time, research concerning the experiences, tasks and employment conditions of migrants working in the global old-age care economy has flourished (e.g., Iecovich, 2010; Karner, 1998; Teshuva et al., 2019; Timonen & Doyle, 2010; Walsh & Shutes, 2013). Part of this literature uses an intersectional approach to unravel the complex interplay of gender, ethnicity and social class in care and employment relationships (e.g., Duffy, 2007; King-Dejardin, 2019; Raghuram, 2019; Williams, 2017). Another strand of literature refers to the global care chain concept (Hochschild, 2000) to explore migrant care workers' life situations and the consequences of their migration for their family members 'left behind' (e.g., Boccagni, 2013; Escrivá & Skinner, 2008; Fudge, 2012; Parreñas, 2010; Skornia, 2014). While the 'global care chain' concept connects the global level of economic inequalities and interdependencies with the local level of (transnational) households, it pays less attention to the multiplicity of actors and institutional intersections also shaping these processes (Michel & Peng, 2017; Yeates, 2009). This is precisely the gap this book seeks to fill by proposing the perspective of the global old-age care *industry*.

2 The Global Old-Age Care *Industry*

Despite notable differences among and within regions and models of old-age care around the globe, a certain convergence across countries can be observed. To keep public expenditure low, old-age care systems have increasingly been opened to the market and steps have been taken to promote (comparatively cheaper) home-based old-age care arrangements (Benería, 2010; Brennan et al., 2012; Theobald & Luppi, 2018). Measures undertaken to achieve this goal include the tightening of eligibility criteria for residential care and the fostering of informal home-based care (de-institutionalization) through cash benefits or paid elder care leave (Da Roit & Moreno-Fuentes, 2019; Maarse & Jeurissen, 2016; Simonazzi, 2009). On the one hand, this process of marketization and re-familization of old-age care has increased the demand for cheap care labor in the private sector of old-age care. On the other hand, the fostering of informal old-age care arrangements has increased the risk of overburdening family carers, leading to a growing demand for market-based solutions, including LIMCs as well as (affordable) nursing homes and outpatient care services. Altogether, the changing public–private 'care mixes' or 'care diamonds' indicate a growing number of actors in the long-term care sector (diversification; Giordano, in this book) and a demand for migrant labor in the old-age care systems of affluent countries (Ogawa, 2018; Razavi, 2017).

The increasing demand for migrant labor in these countries has led to the emergence of what we call a global old-age *industry*. The global old-age care *industry* is not an industry in the usual sense of the production and processing of certain material goods or goods in factories and facilities. It rather encompasses *all actors, institutions and organizations relevant in shaping and conditioning the spatial, organizational, professional and legal dimensions of old-age care provision by migrants across and within national boundaries.* Differently put, it refers to the globalization of old-age care as a multi-actor and multilevel phenomenon in which migrant old-age care workers are channeled (or channel themselves) into different spheres of a country's old-age care system.

The phenomenon of the global old-age care *industry* emerged from what Yeates (2005) has called a 'global care services economy'; other scholars refer to it as a 'global economy of care' (Parreñas, 2005) or 'global market for care labor' (Tronto, 2013). Within this global old-age care *industry*, families, public authorities, non-profit and for-profit care providers increasingly resort to hiring migrant and foreign-born care workers, increasingly competing for this 'resource'. This has been followed by a growing involvement of temporary staffing firms or labor market intermediaries within recruitment schemes, who seek profitability out of the institutionalization of labor migration (Benner, 2003; Coe et al., 2010; Peck & Theodore, 2002). Similarly, commercial brokerage agencies seek to benefit from placing migrant care workers in private households (Elrick & Lewandowska, 2008; Lan, 2016; Leiber et al., 2019). NGOs, transnational activist networks, trade unions and international organizations such as the International Labour Organization and the Organisation for Economic Co-operation and Development are involved, as well as transnational crime organizations engaged in human trafficking (Angeli, 2017; Maroukis, 2017; Palumbo, 2017).

At the same time, the governments of some sending and some receiving countries have become key actors in the global old-age care *industry*. For example, care workers have become 'top exports' for the Philippines, with the Philippine state playing an active brokering role (Fudge & Parrott, 2014; López, 2012). The different roles adopted by states entail essential consequences in terms of the legal status of migrant care workers. Migrant care workers who are recruited for jobs in the formal care sector may or may not have the right to change jobs or to bring over their families (Anderson, 2012; Da Roit & Bochove, 2015; Da Roit & Weicht,

2013; Rostgaard et al., 2011; Theobald, 2011; Van Hooren, 2012). Likewise, LIMCs may or may not have a right to stay and a right to work in the country, and they may be employed in a regular or an irregular manner. In fact, not always are governments interested in regularizing LIMC arrangements (see Horn and Schweppe in this book); other actors may push for certain quality standards and legal certainty (see Leiber and Rossow in this book).

The supranational level may also play an important role. Regional migration regimes may facilitate the recruitment of migrant care workers and reduce the role of national states. The clearest example is the European Union: EU free movement provisions have reduced the possibilities for member states to regulate intra-EU recruitment for unskilled care jobs to practically zero (see Leiber and Rossow in this book), facilitated by the free movement of workers and services. In addition, receiving states may come to agreements with sending states in the region, as illustrated by the example of Japan (also see Ogawa, in this book). Furthermore, the recruitment of care labor also takes place across continents. An example is Germany's "Triple Win program", which recruits care workers for nursing homes from countries like the Philippines, Tunisia and Vietnam. If the sending and receiving countries are relatively close geographically, rotational patterns may develop, enabling the migrant care workers to balance their paid care work abroad with their life situations at home (Chau, 2019; Karakayali, 2010; Satola, 2015).

3 Outline of the Book

This book is organized into three parts. The studies in Part I, 'Policies and Regimes', use multilevel analyses to examine how national policies and regimes influence practices and choices of dependent older people, their families and migrant care workers. Part II, 'Promoters and Intermediaries', brings together studies that focus on particular actors and their role in shaping migrant care work patterns and arrangements. The studies in Part III, 'Risks and Social Protection', focus on risks and vulnerabilities in live-in migrant care arrangements, potential risk and protective factors for the abuse of older care recipients in these care arrangements and sources of social protection for the care workers.

In Part I, the chapters are based on empirical studies in different countries. Ogawa in Chap. 2 focuses on Japan, one of the most rapidly aging countries in the world, with an old-age care sector that is facing severe labor shortages. She considers two different programs through which care workers from Southeast Asian countries are recruited to help solve these shortages: a state-sponsored program and a more recently introduced market-oriented one, in which private for-profit agencies play a central role. Drawing on interviews with various stakeholders in Japan, as well as in sending countries, Ogawa compares how migrants recruited under the two programs have been integrated into the workforce in Japanese old-age care homes. She also compares the long-term impact on the care sector. Her findings indicate that the state-sponsored program offers better opportunities and prospects to the migrant care workers, as well as better guarantees for maintaining the quality of care. It also requires, however, greater investment from the Japanese employers. Care homes with fewer resources are therefore more likely to resort to the market-oriented program, in which a larger share of the costs is passed on to the migrants.

In Japan, migrant care workers are not allowed to work as live-in caregivers in private households. In Chap. 3 by Horn, Schweppe, Böcker and Bruquetas-Callejo, the focus shifts to this type of care arrangement in Germany and the Netherlands, two countries with rather different old-age care regimes. Compared with the Netherlands, the provision of old-age care in Germany has remained much more a family responsibility. This explains why the market for live-in migrant care is much larger in Germany than in its neighbor country. Horn et al. examine whether the different regimes also result in different expectations among family carers who make use of this type of care arrangement. Many studies of live-in migrant care (including Liang in this book) have shown that users are inclined to consider the migrant caregivers as quasi-family, with all the concomitant expectations. Horn et al. examine if users in the Netherlands, due to the country's highly professionalized old-age care regime, are more inclined to compare the migrant caregivers to regular homecare professionals than are their counterparts in Germany. Their findings indicate that this is indeed the case. Their analysis also reveals, however, similarities between family carers in both countries. In both countries, expectations vary and are influenced by characteristics of both the care recipient and the family carer him- or herself.

Giordano in Chap. 4 analyzes three trends in old-age care policies and regimes in Belgium and their micro-level effects. These three trends can be observed in many countries with aging populations: a shift from residential care to home-based care services, the privatization of care services (both residential and home-based care) and the fragmentation and multiplication of actors active in the old-age care sector. Giordano discusses how these trends affect care receivers and their families, as well as female migrants working in this sector. Families must combine and manage various types of homecare services and increasingly resort to private providers and/or hiring care workers on an undeclared basis. There is a growing polarization between families who can and families who cannot afford an optimal mix of care services. The three trends—in combination with families' responses to them—have resulted in poorer working conditions, a growing share of migrant women in the care workforce and care workers being at a higher risk of being discriminated against in the workplace.

Chapter 5 by Weicht concludes this part of the book with some methodological reflections. Weicht observes that many studies argue that one must look at the intersections among various regimes (e.g., care, employment and migration) to explain patterns of migrant care work in different countries. Most existing studies, however, have focused on either micro (individual experiences of migrant care workers) or macro (regime-analysis) perspectives. The relationship between national regimes and migrants' journeys is rarely truly investigated. In this chapter, Weicht analyzes an exemplary empirical case of a foreign-born care worker in Austria in order to investigate if and how individual professional and migration journeys can be used to identify the specific macro and meso arrangements that shape people's ways of entering another country's care labor market and care system.

In Part II, the chapters are based on empirical studies of different actors involved in recruiting migrant workers for care jobs in different European countries. Chapter 6 by Leiber and Rossow and Chap. 7 by Menebröcker focus on labor market intermediaries for live-in migrant care work. Leiber and Rossow examine how private for-profit agencies in Germany and Poland have used new opportunity structures provided by the multilevel (national and EU) regulatory context in which they operate. Legal uncertainty and a lack of enforcement capacity are key features of this regulatory context. Drawing on interviews with heads of agencies and other key actors, Leiber and Rossow analyze how the corporatization of care—that is, for-profit actors entering the care sector and deploying their corporate rationalities and logic—plays out in a largely gray market. Their analysis indicates that to ensure decent care and decent working conditions, state regulation—particularly at the national level—is required because the self-regulation of private companies is insufficient.

Menebröcker focuses on a program set up by Caritas, a non-profit welfare organization in Germany that aims to ensure fair and legal working conditions for migrant live-in carers and decent and safe care for the people in need of live-in care. She has been closely involved in the development of this program and provides a detailed account from an insider perspective. The chapter examines how and why various Caritas branches in Germany—despite critical reactions from other Caritas branches—became involved in the live-in migrant care market, how the CariFair model works and how it is rated by the different parties involved (care recipients and their families, migrant care workers, Caritas staff). The chapter concludes with a brief look at the prospects for solving Germany's old-age care crisis in the longer run by employing migrant workers.

In the following two chapters, the focus shifts to actors involved in the recruitment and incorporation of 'newcomers' in old-age care homes and homecare organizations. Rather than examining cross-border recruitment practices, these chapters focus on the recruitment of migrants who are already present in the country to solve labor shortages in the old-age care sector. Olakivi and Wrede in Chap. 8 analyze how managers of care homes and homecare organizations in Finland morally justify their increased recruitment of foreign-born workers for lower-level old-age care jobs that Finnish-born workers are unwilling to accept. Their analysis identifies three different discursive practices, one of which involves distinguishing 'domestic' from 'international' recruitment. Olakivi and Wrede argue, however, that domestic recruitment can also exploit racialized labor market structures and inequalities.

Chapter 9 by Ham is based on an ethnographic study in a German nursing home that attempted to get refugees and asylum seekers interested in working in the old-age care sector by offering them an orientation and training program. The established care workers in the facility received cultural awareness and sensitivity training. Ham's study contributed to the evaluation of the project by investigating the social dynamics between established workers and newcomers. Her chapter highlights the crucial role of the organizational and institutional context in shaping social processes at the work floor. Both established workers and newcomers felt ignored or disrespected by the facility's management and by society at large. Rather than blurring the boundaries between the two groups, however, factors such as a heavy workload, a lack of resources and a lack of understanding on the part of management reinforced each group's inclination to distance itself from the other group, criticizing the other group's concepts and practices of care, and thereby reinforcing the boundaries between them.

In Part III, the chapters focus on risks and vulnerabilities experienced by care recipients and caregivers in live-in migrant care arrangements, formal and informal sources of social protection for migrant care workers, and potential risk and protective factors for the abuse of older care recipients. Liang in Chap. 10 focuses on the vulnerability of live-in care workers from Southeast Asian countries in Taiwan, another rapidly aging country in Southeast Asia. Liang uses the concept of structural violence to highlight the role of the structural and institutional context in shaping the working conditions, employment relations and well-being of female care workers. In particular, Taiwan's migration policy—and its implementation through private intermediary agencies—places the migrant care workers in a liminal space, where their employers manage and surveil their lives. Drawing on interviews with migrant care workers and employers, Liang describes how both the migrants and their employers tend to focus on their immediate economic interests, and how this prevents them from prioritizing the migrants' health. Like several other chapters in this book (Ogawa; Leiber and Rossow), this chapter argues that the governments of labor-receiving countries have a critical role to play in protecting the rights of migrant live-in care workers, including their right to health.

Österle and Bauer in Chap. 11 explore the experiences and practices of European migrant care workers in accessing and using transnational social protection rights. The chapter focuses on live-in migrant care workers from Central and Eastern European EU member states in Austria.

Austria stands out among other countries where live-in migrant care has become a pillar of the old-age care system because it has made special legal regulations for this type of care arrangement. Under these regulations, most live-in carers are hired as self-employed workers. Building on the European social citizenship literature and drawing on interviews with migrant care workers, Österle and Bauer examine if and how these care workers make use of their formal rights (at the national and EU levels) to social protection. Their findings indicate that in practice, the migrant carers' possibilities to exercise their social rights are constrained by factors such as their rotational migration patterns, the employment mode (self-employment) and the weak regulation of labor market intermediaries in this sector.

Chapter 12 by Van Bochove and Zur Kleinsmiede is based on an exploratory study. It focuses on LIMCs in the Netherlands, the networks of which they are part and their experiences of precariousness. The chapter starts from the assumption that, although live-in care work often puts migrant care workers in precarious positions, this outcome is not self-evident. The chapter aims to explore if and how access to social support may enhance live-in care workers' working conditions and/or help them cope with harsh working conditions. Building on literature regarding care networks and migrants' social networks, and drawing on interviews with live-in migrant care workers and other respondents, the chapter explores the roles of various client-related and non-client-related actors in providing different types of support to LIMCs. The chapter indicates that differences in migrant care workers' access to emotional and practical support may be explained by, among other factors, contract-based differences in rotation patterns.

Finally, Chap. 13 by Horn and Schweppe addresses an issue that so far has largely been ignored within research concerning live-in migrant care arrangements. They focus on abuse and protection against abuse of care recipients in these arrangements and in the context of Germany. Based on a secondary analysis of data gathered in the research project "Emergence and significance of transnational care arrangements (ESTRANCA)", Horn and Schweppe identify potential risks and protective factors. The chapter shows that questions regarding violence and protection against violence in these arrangements must be analyzed in the context of complex, interrelated factors, structures and processes. No clear answers can be given as to whether these arrangements increase or decrease the risk of the abuse and neglect of older dependent people. In addition to the tabooing of abuse in home-based old-age care in general, the problematization of abuse in these particular arrangements is hampered by the common practice of employing around-the-clock migrant carers irregularly.

Ethical issues play an important role in research, particularly in research on more vulnerable groups (e.g., frail and elderly people, irregular migrants). Therefore, all authors have been asked to provide information about the ethical principles guiding their studies. Instead of including them in each chapter, we collected this information and checked it for major ethical issues. We particularly checked that all individuals participating in research had given their consent "knowingly, voluntarily and intelligently, and in a clear and manifest way" (Armiger, 1997: 330). The protection of the individual's anonymity and privacy was another ethical issue particularly considered. This protection is seen as given when no quote or case description can be traced back to a particular individual.

Note

1. The study by Horn and Schweppe (2020) shows that this only partially occurred. Despite the closure of borders, some migrant care workers were still able to cross between countries.

References

Acosta González, E., Picasso Risso, F., & Perrotta González, V. (2018). Cuidados en la vejez en América Latina. Los casos de Chile, Cuba y Uruguay. Konrad-Adenauer-Stiftung e.V., Programa Regional Políticas Sociales en América Latina (SOPLA), Santiago de Chile.

Akemura, S., & Kojima, D. (2018). Japan's long-term care cost projections: Comparison with the European commission ageing report. *Public Policy Review, Policy Research Institute, Ministry of Finance Japan, 14*(4), 541–562.

Ambrosini, M. (2013). *Irregular migration and invisible welfare*. Il Mulino.
Anderson, A. (2012). Europe's care regimes and the role of migrant care workers within them. *Journal of Population Ageing, 5*(2), 135–146.
Anderson, B., & Shutes, I. (2014). *Migration and care labour. Theory, policy and politics* (Migration, Diasporas and Citizenship Series). Palgrave Macmillan.
Anderson, K., & Kvist, E. (2015). The neoliberal turn and the marketization of care: The transformation of eldercare in Sweden. *European Journal of Women's Studies, 22*(3), 274–287.
Angeli, D. (2017). Migrant domestic workers and human trafficking in Greece: Expanding the narrative. *Journal of Immigrant & Refugee Studies, 15*(2), 187–203.
Antanackovic, J., & Bourgeault, I. L. (2013). The employment and recruitment of immigrant care workers in Canada. *Canadian Public Policy, 39*(2), 335–350.
Armiger, B. (1997). Ethics in nursing research: Profile, principles, perspective. *Nursing Research, 26*(5), 330–333.
Asis, M. M. B., & Piper, N. (2008). Researching international labor migration in Asia. *The Sociological Quarterly, 49*(3), 423–444.
Behtoui, A., Boréus, K., Neergaard, A., & Yazdanpanah, S. (2020). Why are care workers from the global south disadvantaged? Inequality and discrimination in Swedish elderly care work. *Ethnic and Racial Studies, 43*(16), 155–174.
Benería, L. (2010). The crisis of care, international migration, and public policy. In *Work and life in the global economy* (pp. 142–164). Palgrave Macmillan.
Benner, C. (2003). *Work in the new economy: Flexible labor markets in Silicon Valley*. Blackwell Publishing.
Boccagni, P. (2013). Caring about migrant care workers: From private obligations to transnational social welfare? *Critical Social Policy, 34*(2), 221–240.
Boyd, M. (2017). Closing the open door? Canada's changing policy for migrant caregivers. In S. Michel & I. Peng (Eds.), *Gender, migration and the work of care* (pp. 167–189). Palgrave Macmillan.
Brennan, D., Cass, B., Himmelweit, S., & Szebehely, M. (2012). The marketisation of care: Rationales and consequences in Nordic and liberal care regimes. *Journal of European Social Policy, 22*(4), 377–391.
Campbell, J. C., & Ikegami, N. (2000). Long-term care insurance comes to Japan. *Health Affairs, 19*(3), 26–39.
Cangiano, A., & Shutes, I. (2010). Ageing, demand for care and the role of migrant care workers in the UK. *Journal of Population Ageing, 3*(1/2), 39–57.

Cangiano, A., & Walsh, K. (2014). Recruitment processes and immigration regulations: The disjointed pathways to employing migrant carers in ageing societies. *Work, Employment and Society, 28*(3), 372–389.

Chan, R. K. H., Soma, N., & Yamashita, J. (2011). Care regimes and responses: East Asian experiences compared. *Journal of Comparative Social Welfare, 27*(2), 175–186.

Chaney, E. (1991). *Muchachas no more: Household workers in Latin America and the Caribbean*. Temple University Press.

Chau, H. S. (2019). Producing (im)mobilities in home care for the elderly: The role of home care agencies in Switzerland. *International Journal of Aging and Later Life, 13*(2), 23–50.

Chau, H. S. (2020). *Brokering circular labour migration: A mobile ethnography of migrant care workers' journey to Switzerland*. Routledge.

Chien, Y. C. (2018). The struggle for recognition: The politics of migrant care worker policies in Taiwan. *Critical Sociology, 44*(7–8), 1147–1161.

Choi, J. C. (2012). End of the era of productivist welfare capitalism? Diverging welfare regimes in East Asia. *Asian Journal of Social Science, 40*(3), 275–294.

CIC NEWS. (2020). Canada provides PR pathway to asylum claimants in healthcare. Retrieved September 2020, from https://www.cicnews.com/2020/08/canada-provides-pr-pathway-to-asylum-claimants-in-healthcare-0815476.html#gs.fmy05t

Coe, N. M., Jones, K., & Ward, K. (2010). The business of temporary staffing: A developing research agenda. *Geography Compass, 4*(8), 1055–1068.

Da Roit, B., & Van Bochove, M. (2015). Migrant care work going Dutch? The emergence of a live-in migrant care market and the restructuring of the Dutch long-term care system. *Social Policy and Administration, 51*(1), 1–19.

Da Roit, B., & Moreno-Fuentes, F. J. (2019). Cash for care and care employment: (Missing) debates and realities. *Social Policy & Administration, 53*(4), 596–611.

Da Roit, B., & Weicht, B. (2013). Migrant care work and care, migration and employment regimes: A fuzzy-set analysis. *Journal of European Social Policy, 23*(5), 469–486.

Degiuli, F. (2007). A job with no boundaries. Home eldercare work in Italy. *European Journal of Women's Studies, 14*(3), 193–2017.

Der Spiegel. (2020, March 24). Engpässe in der Pflege wegen Coronavirus. Die Betreuungskrise. Retrieved September 28, 2020, from https://www.spiegel.de/wirtschaft/soziales/corona-krise-trifft-die-pflege-die-krise-ist-laengst-da-a-a483b3a0-80c0-4a2f-8940-a376801e19cb

Duffy, M. (2007). Doing the dirty work. *Gender & Society, 21*(3), 313–336.

Elrick, T., & Lewandowska, E. (2008). Matching and making labor demand and supply: Agents in Polish migrant networks of domestic elderly care in Germany and Italy. *Journal of Ethnic and Migration Studies, 34*(5), 717–734.

Escrivá, A., & Skinner, E. (2008). Domestic work and transnational care chains in Spain. In H. Lutz (Ed.), *Migration and domestic work* (pp. 113–126). Ashgate.

Esping-Andersen, G. (1999). *Social foundations of postindustrial economies.* Oxford University Press.

Franklin, B., & Brancati, C. U. (2015). Moved to care: The impact of migration on the adult social care workforce. *The International Longevity Center—UK*, London.

Fudge, J. (2012). Global care chains: Transnational migrant care workers. *The International Journal of Comparative Labour Law and Industrial Relations, 28*, 63–69.

Fudge, J., & Parrott, D. (2014). Placing Filipino caregivers in Canadian homes. Regulating transnational employment agencies in British Columbia. In K. Strauss & J. Fudge (Eds.), *Temporary work, agencies and unfree labour. Insecurity in the new world of work* (pp. 70–93). Routledge.

Giordano, C. (2020). Freedom or money? The dilemma of migrant live-in elderly carers in times of COVID-19. *Gender, Work & Organization*. https://doi.org/10.1111/gwao.12509

Hochschild, A. (2000). Global care chains and emotional surplus value. In W. Hutton & A. Giddens (Eds.), *On the edge: Living with global capitalism* (pp. 130–146). Jonathan Cape.

Hondagneu-Sotelo, P. (2001). *Doméstica: Immigrant workers cleaning and caring in the shadows of affluence.* University of California Press.

Hong, S.-I., & Han, C.-K. (2017). A multi-dimensional analysis on care regime of foreign care workers in South Korea. *Asia Pacific Journal of Social Work and Development, 27*(3–4), 159–173.

Horn, V., & Schweppe, C. (2020). Häusliche Pflege in Zeiten von Covid-19. *Nachrichtendienst des Deutschen Vereins für öffentlichen und private Fürsorge e.V.*, October 2020.

Horn, V., Schweppe, C., Böcker, A., & Bruquetas Callejo, M. (2019). Live-in migrant care worker arrangements in Germany and the Netherlands: Motivations and justifications of family decision-making. *International Journal of Ageing and Later Life, 13*(2), 83–114.

Howe, A. L. (2009). Migrant care workers or migrants working in long-term care? A review of Australian experience. *Journal of Aging & Social Policy, 21*(4), 374–392.

Huang, S., Yeoh, B. S. A., & Toyota, M. (2012). Caring for the elderly: The embodied labour of migrant care workers in Singapore. *Global Networks, 12*(2), 195–215.

Ibáñez, Z., & León, M. (2014). Resisting crisis at what cost? Migrant care workers in private households. In *Migration and care labour* (pp. 110–129). Palgrave Macmillan.

Iecovich, E. (2010). Tasks performed by primary caregivers and migrant live-in homecare workers in Israel. *International Journal of Ageing and Later Life, 5*(2), 53–75.

Independent People Homecare Services. (2020). Live-in care and the Coronavirus outbreak. Retrieved September 2020, from https://www.ip-live-in-care.co.uk/news/live-in-care-and-the-coronavirus-outbreak/

Isaksen, L. W. (2010). *Global care work: Gender and migration in nordic societies*. Nordic Academic Press.

Iwagami, M., & Tamiya, N. (2019). The long-term care insurance system in Japan: Past, present, and future. *Japan Medical Association Journal, 2*(1), 67–69.

Jönson, H., & Giertz, A. (2013). Migrant care workers in Swedish elderly and disability care: Are they disadvantaged? *Journal of Ethnic and Migration Studies, 39*(5), 809–825.

Karakayali, J. (2010). *Transnational Haushalten. Biografische Interviews mit care workers aus Osteuropa [Transnational householding. Biographical interviews with care workers from Eastern Europe]*. VS Verlag für Sozialwissenschaften.

Karner, T. X. (1998). Professional caring: Homecare workers as fictive kin. *Journal of Aging Studies, 12*(1), 69–82.

Kim, M. M. S. (2015). *Comparative welfare capitalism in East Asia: Productivist models of social policy*. Palgrave Macmillan.

King-Dejardin, A. (2019). *The social construction of migrant care work. At the intersection of care, migration and gender*. ILO.

Lamura, G., Chiatti, C., Di Rosa, M., Melchiorre, M. G., Barbabella, F., Greco, C., Principi, A., & Santini, S. (2010). Migrant workers in the long-term care sector: Lessons from Italy. *Health and Ageing Newsletter, 22*, 1–6.

Lan, P.-C. (2016). Deferential surrogates and professional others: Recruitment and training of migrant care workers in Taiwan and Japan. *Positions, 24*(1), 253–279.

Lehmann, T. (2020, April 3). Kontrollen und Personalmangel wegen Corona. An der Grenze. Retrieved September 28, 2020, from https://www.spiegel.de/politik/coronavirus-grenzkontrollen-und-personalmangel-a-4b60ad48-9fd3-4f29-8e10-79d6a7edaded

Leiber, S., Matuszczyk, K., & Rossow, V. (2019). Private labor market intermediaries in the Europeanized live-in care market between Germany and Poland: A typology. *Zeitschrift Für Sozialreform, 65*(3), 365–392.

Leiblfinger, M., et al. (2020). Impact of COVID-19 policy responses on live-in care workers in Austria, Germany, and Switzerland. *Journal of Long-Term Care, 2020*, 144–150.

León, M. (2010). Migration and care work in Spain: The domestic sector revisited. *Social Policy and Society, 9*(3), 409–418.

Leutz, W. (2007). Immigration and the elderly: Foreign-born workers in long-term care. *Immigration Policy in Focus, 5*(12), 1–12.

Lewis, M. (1992). The self in self-conscious emotions. *Society of Research in Child Development, 57*(1), 85–95.

Lloyd-Sherlock, P. (2018). Long-term care for older people in South 2 Africa: The enduring legacies of apartheid 3 and HIV/AIDS. *Journal of Social Policy, 48*(1), 147–167.

Lobel, O. (2001). Class and care: The roles of private intermediaries in the in-home care industry in the United States and Israel. *Harvard Journal of Law & Gender, 24*, 89–140.

López, M. (2012). Reconstituting the affective labour of Filipinos as care workers in Japan. *Global Networks, 12*(2), 252–268.

Maarse, H., & Jeurissen, P. (2016). The policy and politics of the 2015 long-term care reform in the Netherlands. *Health Policy, 120*(3), 141–145.

Maroukis, T. (2017). Keeping up appearances: The British public policy response to the trafficking of domestic workers in a changing regime of social protection. *Journal of Immigrant & Refugee Studies, 15*(2), 155–170.

Matus-López, M., & Pedraza, C. C. (2016). New long-term care policies in Latin America: The national system of care in Uruguay. *Journal of the American Medical Directors Association, 17*(7), 1–3.

Michel, S., & Peng, I. (2017). *Gender, migration and the work of care*. Palgrave Macmillan.

Morrison-Dayan, R. (2019). A viable care migration programme for Australia's ageing population: Recruiting migrant personal care workers for Australian aged care. *Australian Journal of Social Issues, 54*(4), 371–385.

Mot, E., Aouragh, A., De Groot, M., & Mannaerts, H. (2010). The Dutch system of long-term care. *ENEPRI Research Report No. 90*, 15 June 2015.

Ogawa, R. (2018). Care and migration regimes in Japan, Taiwan and Korea. In R. Ogawa, R. K. H. Chan, A. Oishi, & L.-R. Wang (Eds.), *Gender, care and migration in East Asia* (pp. 181–204). Palgrave Macmillan.

Olivares-Tirado, P., & Tamiya, N. (2014). *Trends and factors in Japan's long-term care insurance system. Japan's 10-year experience*. Springer VS.

Österle, A., & Bauer, G. (2016). The legalization of rotational 24-hour care work in Austria: Implications for migrant care workers. *Social Politics, 23*(2), 192–213.

Palumbo, L. (2017). Exploiting for care: Trafficking and abuse in domestic work in Italy. *Journal of Immigrant & Refugee Studies, 15*(2), 171–186.

Parreñas, R. S. (2001). *Servants of globalization*. Stanford University Press.

Parreñas, R. S. (2005). *Children of global migration: Transnational families and gendered woes*. Stanford University Press.

Parreñas, R. S. (2010). Transnational mothering: A source of gender conflicts in the family. *North Carolina Law Review, 88*, 1825–1856.

Peck, J., & Theodore, N. (2002). Flexible recession: The temporary staffing industry and mediated work in the United States. *Cambridge Journal of Economics, 31*, 171–192.

Raghuram, P. (2012). Global care, local configurations—Challenges to conceptualizations of care. *Global Networks, 12*(2), 155–174.

Raghuram, P. (2019). Race and feminist care ethics: Intersectionality as method. *Gender, Place & Culture, 26*(1), 1–25.

Razavi, S. (2017). Care going global? Afterword. In S. Michel & I. Peng (Eds.), *Gender, migration and the work of care* (pp. 295–304). Palgrave Macmillan.

Rostgaard, T., Chiatti, C., & Lamura, G. (2011). Care migration: The South–North divide of long-term care. In B. Pfau-Effinger & T. Rostgaard (Eds.), *Care between work and welfare in Europe* (pp. 129–154). Palgrave Macmillan.

Satola, A. (2015). Migration und irreguläre Pflegearbeit in Deutschland. Cinteus Band 14. Fachhochschule Fulda.

Simonazzi, A. (2009). Care regimes and national employment models. *Cambridge Journal of Economics, 33*(2), 211–232.

Skornia, A. K. (2014). *Entangled inequalities in transnational care chains: Practices across the borders of Peru and Italy*. transcript.

Soma, N., Yamashita, J., & Chan, R. K. H. (2011). Comparative framework for care regime analysis in East Asia. *Journal of Comparative Social Welfare, 27*(2), 111–121.

Song, J. (2015). Labour markets, care regimes and foreign care worker policies in East Asia. *Social Policy Administration, 49*(3), 376–393.
Spencer, S., Martin, S., Bourgeault, I. L., & O'Sheam, E. (2010). *The role of migrant care workers in ageing societies: Report on research findings in the United Kingdom, Ireland, Canada and the United States*. International Organization for Migration.
Steiner, J., Prieler, V., Leiblfinger, M., & Benazha, A. (2019). Völlig legal!? Rechtliche Rahmung und Legalitätsnarrative in der 24h-Betreuung in Deutschland, Österreich und der Schweiz. *Österreichische Zeitschrift für Soziologie, 44*(1), 1–19.
Stone, R. I. (2016). The migrant direct care workforce: An international perspective. *Journal of the American Society on Aging, 40*(1), 99–105.
Szebehely, M., & Meager, G. (2018). Nordic eldercare—Weak universalism becoming weaker? *Journal of European Social Policy, 28*(3), 294–308.
Teshuva, K., Cohen-Mansfield, J., Iecovich, E., & Golander, H. (2019). Like one of the family? Understanding relationships between migrant live-in care workers and older care recipients in Israel. *Aging & Society, 39*(7), 1387–1408.
Theobald, H. (2011). Migrant carers in elder care provision: Interaction of policy fields. In B. Pfau-Effinger & T. Rostgaard (Eds.), *Care between work and welfare in European societies. Work and welfare in Europe* (pp. 155–176). Palgrave Macmillan.
Theobald, H., & Luppi, M. (2018). Elderly care in changing societies: Concurrences in divergent care regimes—A comparison of Germany, Sweden and Italy. *Current Sociology, 66*(4), 629–642.
Timonen, V., & Doyle, M. (2010). Caring and collaborating across cultures? Migrant care workers' relationships with care recipients, colleagues and employers. *European Journal of Women Studies, 17*(1), 25–41.
Tronto, J. C. (2013). Democratic caring and global responsibilities for care. Cornell University Press. A Paper prepared for Presentation at the Annual Meeting of the Western Political Science Association, Hollywood, CA, 28–30 March 2013.
United Nations, Department of Economic and Social Affairs, Population Division. (2019). *World population ageing 2019: Highlights* (ST/ESA/SER.A/430).
Van Hooren, F. (2012). Varieties of migrant care work: Comparing patterns of migrant labour in social care. *Journal of European Social Policy, 22*(2), 133–147.
Walsh, K., & Shutes, I. (2013). Care relationships, quality of care and migrant workers caring for older people. *Ageing & Society, 33*(3), 393–420.

Wang, M. S., & Chan, T.-C. (2017). The intersections of the care regime and the migrant care worker policy: The example of Taiwan. *Asia Pacific Journal of Social Work and Development, 27*(3–4), 195–210.

Wilfried Martens Centre for European Studies. (2020, May 29). How does the Corona crisis affect the conditions for migrant workers in the EU? Retrieved September 2020, from https://martenscentre.eu/news/how-does-corona-crisis-affect-conditions-migrant-workers-eu

Williams, F. (2017). Intersections of migrant care work: An overview. In S. Michel & I. Peng (Eds.), *Gender, migration and the work of care* (pp. 23–37). Palgrave Macmillan.

World Health Organization. (2017). *Towards long-term care systems in sub-Saharan Africa*. World Health Organization.

Yeates, N. (2005). Global care chains: A critical introduction. *Global Commission on International Migration, 44*, 1–20.

Yeates, N. (2009). Globalizing care economies and migrant workers: Explorations in global care chains. Palgrave Macmillan.

Yeoh, B. S. A., & Huang, S. (2010). Transnational domestic workers and the negotiation of mobility and work practices in Singapore's home-spaces. *Mobilities, 5*(2), 219–236.

Zallman, L., Finnegan, K. E., Himmelstein, D. U., Touw, S., & Woolhandler, S. (2019). Care for America's elderly and disabled people relies on immigrant labor. *Health Affairs, 38*(6), 919–926.

Part I

Policies and Regimes

2

When Local Meets Global: The Changing Face of Old-Age Care in Japan

Reiko Ogawa

1 Introduction

Scholarship on care and migration has become increasingly important within the social sciences during recent decades, when a burgeoning body of literature has focused on migrant women undertaking reproductive work (Ehrenreich & Hochschild, 2003; Anderson & Shutes, 2014; Michel & Peng, 2017). Several factors have underpinned the structural transformation of affluent societies, including the demographic challenges of an aging population, women's entry into the labor market, the commodification and outsourcing of care to the market, and increasing cross-border mobility for the purpose of undertaking domestic and care work (Parrenas, 2003; Oishi, 2005; Zimmerman et al., 2006). The cross-border mobility of migrants has been propelled by the intersection between financial insecurity in the sending countries and privatized, often marketized forms of care in the receiving countries. The decreasing

R. Ogawa (✉)
Graduate School of Social Sciences, Chiba University, Chiba, Japan
e-mail: reiogawa@chiba-u.jp

V. Horn et al. (eds.), *The Global Old Age Care Industry*,
https://doi.org/10.1007/978-981-16-2237-3_2

capacity of families to provide old-age care creates a large care deficit, both for paid and unpaid work.

Research on migration and care has focused particularly upon the intersection between care and migration policies, as well as how this is articulated and manifested at different levels (Michel & Peng, 2017). In East Asia, migrant care workers are incorporated differently within each country's particular nexus of care and migration regimes, which defines their prospects to stay and work in the host society (Ogawa, 2020). This chapter builds upon existing scholarship, arguing that in addition to care and migration policies, institutional frameworks governing the migration of care workers create different types of entitlement and prospects for migrants to enhance or limit their capabilities within the host society.

Japan is one of the world's fastest-graying societies. Nevertheless, it has been reluctant to accept migrants—particularly those in the 'unskilled' sector—due to fears that they will remain in Japan. How care work should be situated within the dichotomized discourse of highly skilled versus unskilled is contentious, as these distinctions are made arbitrarily and may change over time (Ogawa, 2012a). Despite the prevalence of hard-core conservative politicians, Japan gradually opened up its old-age care labor market to migrant workers due to the existing labor shortage.

The first wave of migrant care workers came under the bilateral Economic Partnership Agreement (EPA) between Southeast Asia and Japan. This came unexpectedly as part of the free trade agreement, and the government was extremely reluctant to accept migrant care workers in the beginning. However, following the positive results of accepting migrant care workers under EPA, the Shinzo Abe administration went on to significantly deregulate the country's migration policy. As of 2020, there are at least four different parallel visa categories that enable migrants to undertake old-age care work (Sasaki & Ogawa, 2019).

This chapter examines the two channels developed in Japan for migrants to work in the old-age care sector: a state-sponsored EPA starting in 2008 and a market-oriented Technical Internship Training Program (TITP) initiative that began to encompass old-age care work in 2017. Juxtaposing these two channels makes clear how the different institutional frameworks enable or constrain migrant care workers, while also showing the long-term impact on the care sector. I begin by examining the first wave of care worker migration to Japan under the government-to-government free trade agreement. In this part I probe the encounter between migrant care workers and Japanese care facilities and discuss the contentious issues that led to the transformation of certain workplace practices. Next, I outline the deregulated policies for migrant care workers and discuss the implications for both the care sector and migrants. Here, I argue that a market-oriented migration system is also embedded within state regulatory power, which must meet two goals: protecting migrants' human rights, while ensuring decent working conditions. Instead of perceiving 'migrant care workers' as a homogenous entity, they are differently constructed according to institutional arrangements and resources, that have to be unpacked.

The chapter focuses on labor migrants who came to Japan to work in the old-age care sector, excluding irregular migrants and resident migrants such as those who came via marriage. The term 'migrant care workers' here refers to migrants who work in old-age residential facilities, because home-based care work, whether dispatched or live-in, is not open to labor migrants in Japan.

This chapter is based on research conducted since 2008, including interviews with 72 Japanese staff and 37 migrant care workers at 27 old-age care residential facilities, some of which I visited multiple times. The care facilities were selected based on snowball sampling. The migrant care workers in these facilities arrived primarily under Economic Partnership Agreements (EPA) because aside from resident migrants, they were the only migrant care workers in old-age care until the revision of law in 2017. Interviews with migrants were conducted both inside and outside the care facilities, including at social gatherings after work, or public forums where they were invited to speak. The contact with migrants was maintained through Social Networking Services (SNS) such as Facebook even after some of them have returned to their home countries. Research was also undertaken in Indonesia, the Philippines, and Vietnam, where I interviewed government officers; teachers at nursing schools; directors, teachers, staff, and migrants who are currently under training in sending agencies; staff of international organizations; and members of civil society, including labor unions. In addition, I participated in various meetings and conferences on migration and care ranging from academic conferences, gatherings of care practitioners, civil society advocacy, and home parties with migrants. The data were first coded to identify the recurring topics. The identified codes were then condensed into larger categories. The final step consisted in the elaboration of the relationship between the categories.

2 Population Aging and Japan's Crisis of Old-Age Care

Japan stands at the forefront of aging societies, with one of the world's longest life expectancies. This rapid speed of aging has occurred hand-in-hand with a drastic decline in the total fertility rate, which was as high as 4.54 in 1947 during the post-war baby boomer generation, but dropped to 1.42 in 2018 (MHLW, 2019a). The demographic crisis of population aging has already reached an alarming stage, with one in three persons above 65 years old (a total of 28.4%) as of 2019 (Statistics Bureau, 2020). The productive age population between ages 15 and 64, which peaked at 87 million in 1995, dropped to 75 million in 2018 and is set to decrease further to 64 million in 2035 (Cabinet Office, 2019).

In order to limit the continuing rise of medical expenditures, Long-Term Care Insurance (LTCI) was established in 2000 to distinguish between medical and social care. LTCI is a public social security program that has been funded through social insurance and tax. It is similar to Germany's LTCI, but also has major differences in accordance with Japan's specific context (Campbell, 2009). LTCI aims to shift care responsibilities from the private sphere of the family to society-at-large. It regulates the financing of care through assessments given by care managers and stipulates the provision of care as defined by the public insurance scheme. Unsurprisingly, the establishment of LTCI led to the rapid expansion of the care labor market—along with skyrocketing expenditures to sustain long-term care. Between 2000 and 2017, the number of older persons eligible to receive care services under LTCI grew from 2.5 to 6.4 million, while the number of care workers in the labor market increased from 549,000 to 1,830,000 (MHLW, 2019b). This was accompanied by rising financial costs for LTCI within the same period, from US $33.2 billion to $99.6 billion (Cabinet Office, 2017).

A national certificate for old-age care work (certified care worker, or *kaigo fukushishi*) was established in 1987, but failed to earn recognition as a professional occupation. The expertise and authority of certified care workers are not clearly defined in the certificate. Moreover, the certificate is not mandatory; one can be employed as an old-age care worker without being certified.[1] The salary of old-age care workers is low compared to other health care occupations. In order for them to be properly remunerated as professional workers, an increase in insurance premiums and tax would be inevitable—but the premium has already more than doubled over the past 20 years.[2] Thus, certified care workers occupy the ambivalent position of being entitled to a national certificate, although their 'expertise' is not explicitly recognized, and their salaries are lower than in other sectors. The younger generation is increasingly not eager to take up care work, and even if they do study the profession in vocational schools, they do not necessarily go on to enter the old-age care sector.[3]

In 2025, when the post-war baby boomer generation will be 75 years old, the government estimates that the shortage of care workers will reach 340,000 (MHLW, 2018a). Foreseeing a labor shortage in the old-age care sector, the government drafted the following five measures to ensure the old-age care workforce: (1) increasing salaries, (2) developing a diverse workforce, (3) improving the retention rate, (4) bringing a positive image to old-age care work, and (5) preparing more thoroughly to welcome migrants (MHLW, 2018b). Policy measures (1) to (4) did not have a major impact, however, due to financial constraints in improving working conditions. In the meantime, the opening up of the old-age care labor market to migrant workers was implemented as a political decision through revision of the immigration law.

3 Migrants in the Care Labor Market: The First Wave

Foreign persons in Japan comprise just 2% of the total population, while there are just 1.66 million labor migrants among a workforce of 68.4 million (E-stat, 2020). The proportion of the migrant labor force varies by sector, with more migrants working in construction and manufacturing than in health care (Kato, 2017). Dependency on labor migrants is increasing in all sectors, and the decline foreseen in the productive age population means that this will most likely accelerate in the coming decades.

Japan has been reluctant to admit migrants other than those in the highly skilled category. The entry of migrant care workers has therefore been a contingency, rather than comprising an official part of migration or labor policies (Ogawa, 2012b). The first wave of migrant care workers arrived via the bilateral Economic Partnership Agreement (EPA). This included the Movement of Natural Persons (MNP), where migrants from Southeast Asia came to work in Japan as nurses and old-age care workers.[4] While ASEAN and the EU have a mutual licensure recognition scheme for professional workers such as doctors or nurses, there is no such system for old-age care workers within Asia. Since there is no similar occupation in the sending countries, old-age care workers under the EPA are recruited mostly from among those with a nursing background.[5] The EPA also stated that migrants must pass the national care work examination—which is given in the Japanese language—as a condition for staying beyond the initial four-year period.[6] Once they pass and become certified, they can renew their residential status indefinitely, bring their families, and change employers freely.

Care work performed within intimate spaces provokes cultural and emotional reactions on how care should be organized, and who is legitimate to provide care (Lan, 2010). In the beginning, some Japanese old-age care workers voiced negative opinions such as 'only Japanese can care for Japanese', implying that care workers must be familiar with the cultural background and lived experience of care receivers in order to provide quality care.[7] Indeed, a knowledge of language and culture has been the most contentious issue in terms of judging whether migrant care workers are capable to deliver. In order to protect the domestic labor market, the Ministry of Health, Labor and Welfare (MHLW) was extremely reluctant to accept migrant workers.[8] Despite such discontent and aversion, the first wave of Indonesian care workers arrived in 2008, followed by Filipinos in 2009 and Vietnamese in 2014.

The EPA provided an opportunity for Japanese old-age care facilities to employ migrants as care workers for the first time. Old-age care facilities are spread out nationwide and are sometimes located in very remote communities. In a country like Japan, where politicians repeatedly claim the ethnic homogeneity of the nation,[9] one would wonder how migrant old-age care workers placed in rural Japan would be received by locals—especially the old-aged. Yet, once the EPA began, migrants were generally welcomed warmly (Ogawa, 2012b). How can we understand the positive acceptance of EPA migrant care workers within the homogeneous landscape of Japan?

As this was the first time to employ foreign staff, many Japanese care facilities were extremely nervous in the beginning. The director of a facility in eastern Japan that employs EPA old-age care workers remarked:

> *I was anxious in the beginning and wondered how the older persons and their families would respond. But when they (EPA old-age care workers) came, they became popular among both. It was great that they came. Usually, the families would direct inquiries to Japanese staff who are fluent in Japanese, but they began asking the EPAs regarding various matters. The families seemed to be supportive and trusting of them.*[10]

This interview encapsulates both the concerns of the care facility and the turning point when they began welcoming the migrants. The director appreciated their contribution and indicated that the workplace has primarily benefitted from their existence—a typical view among the care facilities where I conducted interviews. One of the reasons for the positive acceptance of EPA workers is due to their high educational background and one year of Japanese language training, which are both factors enabling them to jump-start their work. The fact that most EPA workers are trained as nurses and are well-equipped with knowledge of diseases, medicine, and body anatomy are some of the factors contributing to their positive acceptance. Research suggests that 70% of the migrant care workers undertake the same tasks as Japanese staff, including feeding, bathing, toileting, transfer, and oral care, but that some lag behind in tasks requiring Japanese language proficiency, such as making care plans, communication with families, and facility management (JICWELS, 2015). The workplace, which was once dominated by Japanese, has been gradually transformed to accommodate migrant care workers within the onsite teams.

EPA migrant care workers are appreciated not only by old-age care facility managers, but also by their Japanese co-workers. Satoko, a Japanese supervisor of EPA care workers, says that when she heard that Anna (an Indonesian care worker) was returning to her home country, she was extremely disappointed and almost burst into tears. She recalled saying, 'Please don't leave. You shouldn't go. We were always together overcoming difficulties in this ward. I beg you to reconsider your decision.'[11] For Satoko, Anna was not just a junior colleague in the workplace, but also an indispensable partner and friend who shared joy and hardship together during several years working in the care facility.

Yumi, another Japanese care worker who was working with an Indonesian care worker, Ayu, was furious when one Japanese volunteer called Ayu a 'foreigner' (gaijin). 'She is not a foreigner; her name is Ayu', Yumi said.[12] She was offended by the apparently alienating phrase 'gaijin', which literally means 'outside person' and did not represent her relationship with Ayu. The national boundary between Japanese and Indonesian had become blurred, and as the migrants became team members who strived for the same goal of providing quality care, stereotypes such as 'gaijin' no longer held. Yumi respected Ayu because of her positive attitude toward work, and her background in nursing. As the EPAs become an integral part of the old-age care workforce, partnerships began taking place on equal footing—thereby changing Japanese perceptions toward migrant care workers.

Other research also resonates with this positive encounter. In a survey of three care facilities in Tokyo that employ EPA care workers, the response from 102 Japanese care workers to the question 'How do you evaluate EPA care workers?' was as follows: 'very highly appreciate'

(35.2%), 'highly appreciate' (44.1%), 'appreciate' (19.6%), 'don't appreciate' (0%), 'don't appreciate at all' (0%), and 'other' (0.9%). The reasons stated were as follows: 'EPAs (migrant care workers) are industrious and make efforts', 'EPAs are serious and help inspire Japanese staff', 'EPAs are positive and cheerful', and 'I don't look at them as foreigners' (TCSW, 2020).

Situating this new EPA migratory flow within the larger context of migration in Japan, it can be seen as a deviation or anomaly insofar as the way that Japanese see other Asian countries is not always positive—a reflection of the historical trajectory of colonialism and self-orientalism (Iwabuchi, 2002). Historically, Japan situates itself between the West and Asia, belonging to neither, and claiming a superior position vis-à-vis its neighboring countries (Iwabuchi, 2002). This self-orientalism on Japan's part has shed light on how Japan treats migrants who come largely from Asia (Tanaka, 2013). Moreover, scholarship on migration and care shows how the intimate sphere has created various cleavages based on gender, class, ethnicity, and religion (Parrenas, 2003; Lan, 2006; Anderson & Shutes, 2014).

What, then, are the factors contributing to the creation of an environment that enables migrants to work in Japan's care facilities? I argue that the preparation prior to migrants' arrival and continuous support for their work comprise the necessary conditions for their smooth acceptance. Care facilities prepare substantially before employing foreign staff for the first time, such as inviting international students from neighboring communities to give lectures on Southeast Asia so that staff can familiarize themselves with the migrants' home countries. The facilities also explain to the families why they employ EPA migrants, and Japanese staff donate furniture and electric appliances so that the EPAs can smoothly begin their new lives in a foreign land.

When the EPA care workers begin working, their employers are mandated to support their studies by employing Japanese language teachers or assigning staff to teach them so that they can pass the national exam on care work and become certified. Governmental or semi-governmental bodies in both the sending countries and Japan serve as mediating agencies that oversee recruitment, match, monitor, and supervise employers, and they also prepare textbooks to support the migrants' study. As a state-sponsored migration project, the EPA is regulated and transparent compared to the market-oriented migration system, requiring migrants to shoulder no costs. All migration-related fees, including that for one year of Japanese language training, are borne by the Japanese government and employers. Underpinned by bilateral government agreement, the program is trusted by both employers and migrants alike.

The average passing rate for the national care work exam in three countries in 2018 was 46.0% (MHLW, 2018c). Between 2008 and 2019, 4302 EPA migrants arrived and were employed as care workers, but just 3165 were working as of January 2019.[13] This means that 1137 left Japan under one of the following circumstances: (1) prior to taking the national exam, (2) not having passed the national exam, or (3) passing the national exam but still deciding to return.

Not so much has been written about the employers who lost workers after having invested in them considerably (Ogawa, 2015). In the case of EPA care workers, employers invested more than US $18,400 for each individual's education and benefits (either cash or in-kind payments) on top of their salaries (Tsubota, 2018). Consequently, employers and Japanese co-workers like Satoko, who befriended the migrant care workers, have felt very disappointed when they left—especially if they passed the national exam. Considerable disappointment has been voiced by employers when the EPA care workers have resigned and returned, particularly those who are popular among the old-aged and the Japanese staff. Migrant care workers under the EPA are neither cheap nor disposable labor, and since migrants did not shoulder any placement fee in coming to Japan, they are in the category of free laborers.

4 Transforming the Workplace: From Contention to Inclusion

While EPA care workers have been well-received, this does not mean that there were no tensions—as is made clear by the large number of returnees. Care Facility S, a large silver industry conglomerate located in a depopulated rural area of western Japan, has been actively employing

EPA workers since the beginning of the program.[14] The facility accepted 148 EPA workers from 2008 to 2013, among which 84 passed the national exam. By 2017, however, just 44 of those who passed the exam were still working there. In the beginning, management was stringent and did not accommodate the migrants' needs, insisting on doing things their own way. Consequently, even though the EPA care workers were well-supported in their studies, they understandably disliked some of the workplace rules—particularly the one forbidding them to use their paid holidays to return to their home countries until they passed the national exam. Ten-day statutory paid holidays are given after six months, with an increase every year. Workers are entitled to decide when and how to use them, but Japanese care workers usually do not take long consecutive holidays.

Long working hours, and overtime work that prevents work-life balance, have been defining characteristics of Japanese corporations. The hierarchical workplace culture suppresses and silences workers from negotiating labor conditions—particularly new, young women workers. Migrant care workers were extremely frustrated, since unless their employers allowed them to take consecutive paid holidays, those from Southeast Asia could not return home—especially since many are not from the capital cities, which would require additional traveling time.

One Indonesian care worker who passed the national exam and was working for five years in Japan had a baby born in West Java, Indonesia. He asked for a two-week leave to be with his family, but his request was denied since two weeks was considered 'too long' and 'unfair' to the Japanese staff. Shocked by this response, he decided to resign even though he was very well integrated within the workplace and community. Imposing exactly the same rules as those applying to Japanese staff—which are, moreover, often not explicit—frustrates the migrants, particularly those who have left their families behind.

The other rule was not allowing Muslim women to wear the veil in the workplace. The management of care facility S cited different reasons, such as 'it frightens the dementia patients', 'older persons may pull it, and it can be dangerous', or 'it's unsanitary'. Some (but not all) Indonesian migrants contested this as non-legitimate reasoning, however. In fact, the old-aged did not care much or make a big deal out of it, especially in rural areas, as the veil resembles the scarf worn in farming fields to avoid the heat and wind.

There are more than 60,000 Indonesians in Japan, comprising the country's largest Muslim population. Except for EPA care workers, however, most of them are either students or production sector employees. The first mosque was built in Kobe in 1935, but historical encounters with Muslims are far more recent compared to Europe. No state policies exist in Japan to explicitly intervene in order to decide which religious symbols are permissible or not, including the Muslim veil (Foner & Alba, 2008). Without much knowledge on religion in general, and Islam in particular, the Muslim veil was never associated with connotations such as women's oppression, limitation of freedom, or going against modernity (Scott, 2007). Rather, it has translated into something that rural folks are familiar with, that is, a scarf to be worn in the fields. The old-aged in the care facility wish for the migrants to stay and provide care work as long as they are kind, gentle, cheerful, and loving; the veil is not an issue.

Care facilities have not been sensitive enough toward migrants' religious needs; however, so many EPA care workers resigned or were forced to resign—including those who had successfully moved to different care facilities. The asymmetrical relationship between employer and employee served to silence migrants. Even though some tried hard to negotiate, it generally ended in failure.

Quite a number of EPAs have returned home, but there is no comprehensive data on the reasons for returning. Most returnees indicated 'family' as an 'official' reason, such as needing to care for grandparents or parents, or getting married. Beneath the surface, however, their decisions were impacted by power relations, physically demanding work, and a rigid culture within the care facilities. Workplace bullying, discrimination and harassment, restrictions on wearing the Muslim veil, minimal entertainment in rural areas, and lacking future prospects are some of the reasons migrant care workers gave for their decision to return home.[15]

Migration scholarship, especially the existing literature on migrant domestic workers, examines their subjective experience, which includes vulnerability, displacement, and control (Anderson, 2000; Constable, 2007). While an asymmetrical relationship persists between employer and migrants, or between native and migrant workers, 'free labor' migrants are in a position to envision and compare other opportunities. And with the economic developments taking place in Southeast Asia, migrant care workers also likely have better options in their home countries than staying in Japan as care workers.[16]

Losing numerous EPAs after building a new dormitory and investing in their education, the management of care facility S finally understood the need to review their policies to facilitate retention. Today, the EPA care workers can use their paid holidays to return home, and after the resignation of many EPA Indonesian care workers, Muslims may now wear veils at the care facility S.[17]

Through the EPA, many care facilities learned to accommodate Muslim employees' religious practices, such as preparing a prayer room and securing time for prayers, cultivating understanding toward Halal food, and exempting them from physically demanding work during Ramadan. Amidst a competitive environment for finding care workers, such facilities are successful in attracting the migrants through word of mouth. EPA care workers have exercised agency in negotiating with their employers and have leveraged themselves by gaining trust in the workplace.

To encourage retention, care facility S began introducing a family allowance for those who passed the exam and brought their families, along with free kindergarten for kids and free Japanese language classes for spouses. While providing incentives to the migrant care workers, other care facilities are also making efforts to mitigate tensions and discrimination between Japanese staff and migrant care workers. An example at one facility is an annual training program in the Philippines for Japanese staff, which allows them to imagine and experience what it is like to live and work in a foreign country. To avoid differences such as ethnicity, class, gender, or religion being translated into markers of social divisions, such efforts greatly help cultivate shared values while integrating migrants into the workplace.

Migrant care workers have challenged Japan's homogeneous workplace and working culture, creating new models that have encouraged sensitivity and flexibility toward different ideas and practices. After ten years, many EPAs who passed the certification exam and stayed in Japan have become leaders and managers tasked with supervising Japanese staff.

Some, aspiring toward upward mobility, have even passed the higher qualification exams to become care managers or nurses. Following certification, they can change employers, bring their families to Japan, and renew their visas indefinitely. As of June 2019, 431 relatives are living in Japan under EPA family visas, including spouses and children, and some have already obtained permanent residency. Their citizenship status entitles them to freedom of labor, family reunions, and increased life options to move between transnational spaces. The care facilities employing EPAs have realized that simply imposing the hierarchical and rigid Japanese management style does not work, and they have become more sensitive toward respecting diversity and attending to migrant care workers' needs. EPA care workers will continue to work if their religious, spiritual, material, and family needs are respected. Moreover, this change in workplace rules and regulations has been beneficial not only for the migrants—but also for Japanese.

Being a very expensive program with high requirements, however, the number of EPA care workers remains minimal after ten years of implementation and is still too limited to mitigate existing labor shortages. In addition, the care facilities are too burdened and overstretched to support EPA care workers with the Japanese language and national exam preparations. Although the state provides textbooks, curricula, and trainings, the daily study is overseen by the care facilities, which are already overwhelmed by labor shortages.

5 Deregulation of Care Worker Migration: Three More Waves

Foreseeing that the needs associated with the rapid speed of aging cannot be met with the small number of EPAs, the Abe administration introduced two new channels in 2017 to accept migrant workers in the old-age care sector, and one more channel in 2019. First, a new visa category was created called 'long-term care work' (kaigo), which targeted international students who graduate from vocational caregiving schools. Second, old-age care was added to the Technical Internship Trainee Program's

(TITP) list of occupations. Finally, the new Specified Skilled Worker (SSW) visa category was created, targeting labor migration in 14 occupations such as agriculture, construction, and old-age care. For SSWs, the government set the maximum number of migrant workers within the next five years at 340,000, including 60,000 care workers.

Aside from the EPA, these three schemes, which run parallel to each other, are not coordinated on the basis of career development, which would ensure that migrants in all channels would receive increased remuneration as they advance their skills and experiences. One reason why this complex framework did not draw much criticism locally is the positive acceptance of EPA care workers amidst the ongoing labor shortage within the care sector. Except for the EPA, which is administered by the state and does not shoulder migrants with any costs, the three new channels are mediated by private agencies, and the payment infrastructure is not transparent (Sasaki & Ogawa, 2019). This neo-liberal turn has emphasized economic efficiency and cost reduction, and the new migratory flows will be dominated by the market through private brokers. However, it should be noted that a juxtaposition between the state-led EPA and the market-led TITP would in fact represent an 'ideal type', as TITP also features a bilateral MOU with the sending countries, as well as oversight by a governmental body, while the migration process is mediated by private agencies. Therefore, the market-oriented migration system is still within the realm of state regulatory power (Xian, 2012).

Among the three new channels, TITP has a longer history and is comparable with the preceding EPA, while SSW is about to start. Moreover, space here is too limited to discuss student migration. This section will provide an overview of TITP, comparing the institutional framework of TITP old-age care worker migration with EPAs.

TITP was conceived in 1993 to 'train' migrant workers for three years in 'skill transfer', which is meant to contribute to the development of their countries. Occupations such as agriculture, manufacturing, food processing, and construction were included, and the aspect of 'training' has been emphasized rather than that of 'labor'. This is because the state believed that (a) unskilled labor migration should not be allowed, (b) migrants should not take away jobs from native workers, and (c) migrants should not be permitted to settle (Kamibayashi, 2015). The period of stay is limited to five years, and migrants' status can be changed from TITP to SSW. Altogether, the maximum period of stay is 10 years, with no possibility of renewal or family reunion unless the national exam on care work is passed.

TITP migrants receive limited language training in the sending countries and are not allowed to change their employers, so their bargaining power is far less than the EPA migrants. Private agencies on both sides will strictly monitor and control the migrants to silence them, sometimes in an oppressive and disciplinary manner.[18] Under the privatized migration channel with brokers on both sides, migrants (particularly those from Vietnam) come with debt bondage, and their labor rights have been significantly restricted. Although they enter the country regularly via the TITP program, once in the country they overstay and become undocumented, which results in an increasing number of undocumented migrants (Sunai, 2019). Consequently, the civil society has repeatedly called for termination of the program.[19] Instead, however, the government decided to expand existing job categories to include old-age care work within the occupational list. Unlike other occupations without language requirements, however, Japanese language ability became a condition for the TITP old-age care workers.[20]

While EPA is a government-to-government program, TITP's migration system in both the sending countries and Japan is mediated by the market under state regulatory power, which tries to achieve two often conflicting goals: protecting migrants' human rights and bringing profit to private agencies. The Philippines has the most progressive migration policy among the three sending countries; migrants can be charged with no placement fee (Ogawa & Sadamatsu, 2020). The Vietnamese government sets the maximum placement fee at US $3600, but in reality, migrants pay approximately US $9200, which they borrow in Vietnam (Sunai, 2019, p. 59). In Indonesia, the TITP migrants are charged between US $3000 and $4000, with an overall predeparture process that is not transparent (HRWG, 2020).

The old-age care facilities are confused by the complexity of the new channels, but they have had no choice except to sort through the different available options. For EPA, the eligibility for care facilities to employ migrants is stringent, making it impossible for small, resource-challenged facilities to employ EPA care workers, because they need to provide support for migrants to pass the national exam. For such facilities, hiring TITP old-age care workers through private agencies has been easier than employing EPAs. However, without appropriate knowledge about the complexity of migration laws and bureaucracy, they had to completely depend upon private agencies.

As of February 2020, more than 2500 Japanese license supervising agencies are dispatching TITP workers to Japanese companies, including 700 that claim to dispatch care workers from countries such as China, Vietnam, the Philippines, Myanmar, and Indonesia (OTIT, 2020). As of March 2019, a total of 1823 old-age care workers had arrived among the 389,321 total TITPs (OTIT, 2019).

Although Vietnam is considered a major country in terms of sending migrant workers to Japan, mixed opinions were heard from the Vietnamese side regarding the newly established care worker program. The director of a sending agency in Hanoi that has been sending large numbers of TITPs in the production sector was cautious regarding the dispatching of care workers:

> *Care work is different from other types of work, as it is difficult to foresee what kind of problems will occur. For example, EPA old-age care workers experienced groping (sexual harassment). Care is meant to support human lives. You are not allowed to make a single mistake like breaking something in a factory, so high language proficiency is required. However, this means that the migrants must be trained for a longer period, which makes the program unattractive.*[21]

His remark succinctly summarizes the nature of care work to be provided in the intimate sphere, which requires certain skills and training, but is subject to less monetary remuneration. The average salary for workers in Japan in regular employment among all sectors is 249,500 yen (US $2287) (25–29 years old, before tax deduction) and 198,900 yen (US $1823) for non-regular workers of the same age. For TITP workers, however, it is 156,900 yen (US $1438) (26.7 years old working for 1.5 years), which is 60% of the Japanese salary (MHLW, 2019c). For old-age care workers in institutional care, the monthly salary before tax deduction is 202,145 yen (US $1853), which is lower than the average of all sectors

(Kaigo Rodo Antei Center, 2018). How much, then, can a TITP migrant old-age care worker expect to earn?

A well-established sending agency in Jakarta with a long history of dispatching TITPs says that for Indonesian TITP care workers, the salary after deducting tax and social insurance is 90,000 yen (US $825) for the first year, 95,000 yen (US $870) the second year, and 100,000 yen (US $916) the third year.[22] In principle, there should be no discrepancy in salary between Japanese and migrant workers, but a gap does exist in practice. From this, we may understand that if the migrant's salary is less than that of Japanese workers, migrants who enter the care sector—which is already low paid—will earn even less. From the migrants' viewpoint, then, working in a factory is a better option, since language requirements are lower and salaries higher.

The language requirement for care work prolongs the migrants' training period and increases the already exorbitant recruitment fee, while neither ensuring government support nor permitting access to permanent residency or Japanese citizenship. It is likely that migrants become constrained with the debt accumulated during their prolonged training and struggle with limited language abilities that make it hard for them to adapt to the workplace. Since the return process is embedded within the migration trajectory of TITP, capital accumulation is limited—thereby forcing migrants to become undocumented (Belanger et al., 2011).

The TITP has been severely criticized, both nationally and internationally, as a de facto import channel for 'cheap labor' rather than providing technical training to less affluent countries (Kamibayashi, 2015; Sunai, 2019). Since there is no job market in the sending countries for migrants to become employed as old-age care workers, the TITPs rational of 'skill transfer' does not hold.

A number of interviews reveal that TITP migrants are treated differently from those under EPAs, as the TITP workers receive no state support. Care facility S, whose EPA old-age care workers have a good track record of passing the national exam, says that unlike under the EPA, there is no government subsidy to support TITP workers' study—so they will not 'invest' in the training. The salary will also be different, although migrants under the two different channels may come from the same country with similar backgrounds. Currently, private agencies are conducting recruitment in nursing schools in Southeast Asia, but differing treatment in the workplace and without a prospect for career development and citizenship, it will bring a sense of relative deprivation to some of these migrants.

An ex-EPA care worker compares her advantageous position in the care facility vis-à-vis her TITP colleague. The care facility invested in her study through state subsidy and will buy anything she needs to pass the national exam, including textbooks and a dictionary. Her TITP junior colleague receives no such support, however, so she decided to leave behind some textbooks when she resigned. Apparently, the managers of the care facilities will not take time or spend money on migrants who are time-bound.

The shift in migration regimes from EPA to TITP and other channels entails decentralization of power and authority in managing migration. The state's central responsibility to protect migrant's rights and ensure decent working conditions, even in rhetoric, will become diffused as full oversight over the migration trajectory is no longer the sole responsibility of the state. Under this neo-liberal turn, the tasks of recruitment, training, matching, deployment, and monitoring will be conducted by a large number of private agencies, and the state will depend on the market to supply migrant care workers. The fragmented migration policy will affect skill development, while conferring advantages upon those in intermediary positions such as private agencies. Meanwhile, however, the responsibility to provide care will remain in the hands of the care facilities, who pay an expensive placement fee to the private agencies, while simultaneously shouldering all risks associated with migrant deregulation, including accidents due to inappropriate training and lack of language proficiency. The Japanese staff will be burdened as preceptors for less-trained migrant care workers, in what is already a busy workplace. Eventually, the old-aged and their families may have to pay the price of deregulation if accidents should occur, which may in turn lead to the demonization of migrant care workers. In the long run, then, the quality of care will become very different from that of EPA care workers, and the migrants will become stratified not by education, merit, or effort—but merely by virtue of the channel through which they have entered Japan.

6 Conclusion

The debate on migration in Japan is predominantly centered around the 'shortage of labor', without addressing the long-term potential impact on the labor market and social integration. Comparing the two migration systems implemented during the past decade, we see that the rules and resources for EPA and TITP are designed differently—thereby either enabling or constraining the migrant care workers. EPA migrants are leveraging themselves by developing trust and becoming an integral part of the care workforce through continuous support from the workplace. I argue that it is the institutional framework of the state-market relationship that in turn defines the entitlement of migrants. Appropriate measures, such as language training and support for career development, serve as the necessary precondition for migrants to enhance their capacity. This in turn allows care facilities to employ quality care workers who will become future leaders. Access to citizenship and family reunion are necessary conditions for migrants to settle and strive for upward mobility, which will in turn maintain the quality of care work.

The two models—state-sponsored EPA and market-oriented TITP—both have their shortcomings, and neither is likely to be sustainable on its own. The EPA is too limited in terms of numbers, too expensive, and too much of a burden for care facilities to support migrants in passing the national exam. Meanwhile, the TITP—with its limited training and low remuneration—will downgrade care work and continue to raise human rights concerns over exploitative agencies and employers. The private agencies that mediate the TITP are also embedded within the state regulatory framework, which affects diplomatic relationships and economic activities between Japan and Southeast Asia.

Instead of a system that differentiates between Japanese and migrants and stratifies migrants systematically, the new system should be based upon two fundamental principles: protection of migrants' rights and ensuring decent working conditions for old-age care workers regardless of their nationality. The policy for old-age care workers should be enabling and have a clear career path accompanied by remuneration and access to citizenship. COVID-19 has drastically altered the way we think of old-age care and essential workers, confirming that migration policy should be designed to do justice both to older persons in Japan and to the country's migrant care workers.

Acknowledgments This research has been funded by JSPS Kakenhi 19K2127 (Reiko Ogawa), Economic Research Institute for ASEAN and East Asia (Yuko Hirano), and the Japan Center for Economic Research (Reiko Ogawa).

Notes

1. In the category of institutional care, just 44% of care workers are certified (MHLW, 2010).
2. Insurance payments are required for those who are above 40 years old. When the LTCI started in 2000, the national average for monthly payments was 2911 yen (US $27). By 2018, it had risen to 6771 yen (US $62) (MHLW, n.d.).
3. Caregiver schools cannot fill their student slots, and enrollment is decreasing every year. The percentage of new enrolled students decreased from 57% in 2014 to 44% in 2018 (Inoue, 2018). Among 6564 graduates of caregiving schools in 2018, 5656 entered the old-age care labor market (Information obtained from Nihon Kaigofukushishi Yosei Shisetsu Kyokai 2020).
4. Although nurses and care workers are included in the EPA provisions, this chapter focuses only on old-age care workers, since these two occupations are different in terms of education, expertise, and the labor market.
5. The recruitment of nurses from countries that lack medical professionals invokes ethical concerns—a matter that is hardly discussed in Japan. Interview with the Ministry of Health in Indonesia in February 2020. For research on EPA returnees, see Kuruniati et al., 2017.
6. Three years of experience are required to take the exam, which migrants take in the fourth year. Technically, they can take the exam twice.
7. Interview with a representative of an old-age care worker association in 2011.
8. Interview with an MHLW government official in 2018.
9. Deputy Prime Minister and Finance Minister Taro Aso said that Japan has been unified under a single language and ethnic identity for the past 2000 years. He later apologized, following strong criticism that such

commentary neglects the diversity of the nation (*Japan Times*, 14 January 2020).
10. Interviewed in March 2019 in Eastern Japan.
11. Interviewed in March 2019 in Eastern Japan.
12. Interviewed in July 2019 in Eastern Japan.
13. Information obtained from the MHLW.
14. Interviewed in July 2011 and March 2018 in Western Japan.
15. Interviews with 12 migrant care worker returnees in Indonesia, the Philippines, and Vietnam.
16. Interviews with 12 migrant care worker returnees revealed that they are working as teachers, instructors, interpreters at Japanese companies, nurses, nurse assistants, call center employees, or self-employed workers. Some are also unemployed.
17. This is still not a general rule for all care facilities, however.
18. Observed during fieldwork in the Philippines, Vietnam, and Indonesia during 2015–2020.
19. The Japan Federation of Bar Association (JFBA) has repeatedly submitted a statement to abolish the TITP, while the Solidarity Network for Migrants in Japan (SMJ), an umbrella organization supporting migrants, has been negotiating with the government twice a year on human rights violations under TITP. This issue has also been covered by journalists (Sunai, 2019).
20. This refers to level 4 of the Japanese Language Proficiency Test (JLPT) at the time of entry (ability to understand basic Japanese).
21. Interviewed in March 2018 in Hanoi.
22. Interviewed in February 2020 in Jakarta. The interview was conducted together with A. Sadamatsu, K. Hirano, and W.H. Istiqomah.

References

Anderson, B. (2000). *Doing the dirty work? The global politics of domestic labour.* Zed Books.

Anderson, B., & Shutes, I. (Eds.). (2014). *Migration and care labour: Theory, policy and politics.* Palgrave Macmillan.

Belanger, D., Ueno, K., Hong, K. T., & Ochiai, E. (2011). From foreign trainees to unauthorized workers: Vietnamese migrant workers in Japan. *Asian and Pacific Migration Journal, 20*(1), 31–53.

Cabinet Office. (2017). Kaigo kyufu to hokenryo no suii. Retrieved February 20, 2020, from https://www5.cao.go.jp/keizai-shimon/kaigi/special/reform/wg1/291108/shiryou1-8.pdf

Cabinet Office. (2019). *Korei Shakai Hakusho*. Retrieved February 18, 2020, from https://www8.cao.go.jp/kourei/whitepaper/w-2019/html/zenbun/s1_1_1.html

Campbell, J. C. (2009). How policies differ: Long-term-care insurance in Japan and Germany (in Japanese). *Shakai Kagaku Kenkyu, 60*(2), 249–277.

Constable, N. (2007). *Maid to order in Hong Kong: Stories of migrant workers*. Cornell University Press.

Ehrenreich, B., & Hochschild, A. (Eds.). (2003). *Global woman: Nannies, maids and sex workers in the new economy*. Metropolitan Books.

E-stat. (2020). *Rodo Ryoku Chosa*. Retrieved December 12, 2019, from https://www.stat.go.jp/data/roudou/index.html

Foner, N., & Alba, R. (2008). Immigrant religion in the US and Western Europe: Bridge or barrier to inclusion? *International Migration Review, 42*(2), 360–392.

Human Rights Working Group (HRWG). (2020). *Shifting the paradigm of Indonesia-Japan Labor Migration Corporation: Voices of returned migrant workers on predeparture processes under EPA and TITP*. HRWG.

Inoue, Y. (2018). *Gaikokujin ryugakusei ni taisuru kyokai no taio*. Presentation at Kaiyokyo Heisei 30 Nendo Zenkoku Kyoshokuin Kenshu Kai, Miyazaki Shimin Hall. Retrieved May 20, 2020, from http://kaiyokyo.net/member/20181227_ryuugakusei.pdf

Iwabuchi, K. (2002). *Recentering globalization: Popular culture and Japanese transnationalism*. Duke University Press.

JICWELS. (2015). *EPA Kaigofukushishi no teichaku sokushin no kadai ni kakawaru chosa hokokusho*. Kokusai Kosei Jigyodan.

Kaigo Rodo Antei Center. (2018). *Kaigo Rodo Jitai Chosa*. Retrieved April 10, 2020, from http://www.kaigo-center.or.jp/report/2019_chousa_01.html

Kamibayashi, C. (2015). *Gaikokujin rodosha ukeire to Nihon shakai: Gino jitshu seido no tenkai to jirenma*. Tokyo University Press.

Kato, M. (2017). Takamaru 'gaikokujin izondo' to kawaranai gaikokujin rodosha no kozo. *Huffpost*, 21 June, Retrieved September 10, 2019, from https://www.huffingtonpost.jp/makoto-kato/foreign_workers_b_17237588.html

Kuruniati, A., Chen, M. C., Efendi, F., & Ogawa, R. (2017). A deskilling and challenging journey: The lived experience of Indonesian nurse returnees. *International Nursing Review,* 64(4), 494–501.

Lan, P. C. (2006). *Global Cinderellas: Migrant domestics and newly rich employers in Taiwan*. Duke University Press.

Lan, P. C. (2010). Cultures of carework, carework across cultures. In L. Grindstaff, M. C. Lo, & J. R. Hall (Eds.), *Handbook of cultural sociology*. Routledge.

Ministry of Health, Labour and Welfare (MHLW). (2010). Kaigo shokuin ni shimeru kaigo fukushishi no wariai ni tsuite. Retrieved April 2, 2020, from https://www.mhlw.go.jp/stf/shingi/2r9852000000zdft-att/2r9852000000zdr9.pdf

MHLW. (2018a). Dai 7 ki kaigo hoken jigyo keikaku ni motozuku kaigo jinzai no hitsuyo su ni tsuite. Retrieved March 30, 2020, from https://www.mhlw.go.jp/stf/houdou/0000207323.html

MHLW. (2018b). Kaigo fukushi jinzai no taisaku to. Retrieved March 30, 2020, from https://www.mhlw.go.jp/topics/2018/01/dl/tp0115-s01-01-05.pdf

MHLW. (2018c). Dai 31 kai kaigo fukushishi kokka shiken ni okeru EPA kaigo fukushi shi kohosha no shiken kekka. Retrieved February 8, 2020, from https://www.mhlw.go.jp/content/12004000/000493552.pdf

MHLW. (2019a). 2019 Jinko dotai tokei no nenkan suikei. Retrieved November 18, 2019, from https://www.mhlw.go.jp/toukei/saikin/hw/jinkou/suikei19/dl/2019suikei.pdf

MHLW. (2019b). Kaigo rodo no genjo. Retrieved February 14, 2020, from https://www.mhlw.go.jp/content/12602000/000482541.pdf

MHLW. (2019c). Chingin kozo kihon tokei chosa. Retrieved April 10, 2020, from https://www.mhlw.go.jp/toukei/itiran/roudou/chingin/kouzou/z2019/index.html

MHLW. (n.d.). Kaigo hoken zaisei. Retrieved March 20, 2020, from https://www.mhlw.go.jp/stf/seisakunitsuite/bunya/hukushi_kaigo/kaigo_koureisha/zaisei/index.html

Michel, S., & Peng, I. (Eds.). (2017). *Gender, migration and the work of care: A multi-scalar approach to the Pacific Rim*. Palgrave Macmillan.

Ogawa, R. (2012a). Conceptualizing transnational migration of care workers: Between 'skilled' and 'unskilled'. *ASIEN: The German Journal on Contemporary Asia, 124*, 95–114.

Ogawa, R. (2012b). Globalization of care and the context of reception of Southeast Asian care workers in Japan. *Southeast Asian Studies, 49*(4), 570–593.

Ogawa, R. (2015). Demographic change and migration of care workers: state, care facilities and migrants. In S. N. Nagy (Ed.), *Japan's demographic revival: Rethinking migration, identity and sociocultural norms*. World Scientific Publishing.

Ogawa, R. (2020). Making migrant care workers in East Asia. In J. Liu & J. Yamashita (Eds.), *Routledge handbook of East Asian gender studies* (pp. 123–143). Routledge.

Ogawa, R., & Sadamatsu, A. (2020). Reality regarding the institutionalization of Specified Skilled Workers: Findings from the survey on migrant care workers in the Philippines and Vietnam (in Japanese). *Imin Seisaku Kenkyu, 12*, 28–48.

Oishi, N. (2005). *Women in motion: Globalization, state policies, and labor migration in Asia.* Stanford University Press.

Organization for Technical Intern Training (OTIT). (2019). FY 2019 statistics. https://www.otit.go.jp/research_toukei_h29h30/

OTIT. (2020). Search for Japanese supervising organizations. Retrieved February 3, 2020, from https://www.otit.go.jp/search_kanri/

Parrenas, R. S. (2003). *Servants of globalization: Women, migration, and domestic work.* Ateneo de Manila Press.

Sasaki, A., & Ogawa, R. (2019). Social stratification of migrant care workers in Japan. *ASEAN Social Work Journal, 7*(1), 15–34.

Scott, J. W. (2007). *The politics of the veil.* Princeton University Press.

Statistics Bureau. (2020). Tokei kara mita waga kuni no koreisha. Retrieved May 25, 2020, from https://www.stat.go.jp/data/topics/topi1210.html

Sunai, N. (2019). *Dorei Rodo.* Kaden Sha.

Tanaka, H. (2013). *Zainichi gaikokujin: Ho no kabe, kokoro no kabe.* Iwanami Shinsho.

Tokyo Council of Social Welfare (TCSW). (2020). *Gaikokujin no Kaigo Jujisha no Ukeire Gaido Book.* https://www.tcsw.tvac.or.jp/bukai/kourei/documents/gaikokujin.pdf

Tsubota, K. (2018). Who pays the cost and who receives the benefit? Comparing migration policies for care workers in Japan and Taiwan. In R. Ogawa, R. K. H. Chan, A. Oishi, & L. R. Wang (Eds.), *Gender, care and migration in East Asia* (pp. 157–179). Palgrave Macmillan.

Xian, B. (2012). Predatory princes and princely peddlers: The state and international labour migration intermediaries in China. *Pacific Affairs, 85*(1), 47–68.

Zimmerman, M., Litt, J., & Bose, C. (2006). *Global dimension of gender and care work.* Stanford University Press.

3

Family Carers' Expectations and Strategies in Shaping Live-in Migrant Carer Arrangements: A Comparison Between Germany and the Netherlands

Vincent Horn, Cornelia Schweppe, Anita Böcker, and María Bruquetas-Callejo

1 Introduction

Care arrangements with migrants providing around-the-clock care and supervision for older people in private households have become a vital part of long-term care (LTC) regimes in welfare systems labeled as familistic or conservative. In Europe, live-in migrant carer (LIMC) arrangements are particularly common in countries like Italy and Spain and to a

V. Horn (✉) • C. Schweppe
Institute of Education, Johannes Gutenberg University of Mainz, Mainz, Germany
e-mail: hornv@uni-mainz.de; c.schweppe@uni-mainz.de

A. Böcker
Faculty of Law, Radboud University Nijmegen, Nijmegen, The Netherlands
e-mail: a.bocker@jur.ru.nl

V. Horn et al. (eds.), *The Global Old Age Care Industry*,
https://doi.org/10.1007/978-981-16-2237-3_3

lesser extent in Germany and Austria. Their emergence has been explained by social and demographic changes (especially changing family structures, gender roles and increased old-age care needs) and the lack of adequate policy responses and public LTC infrastructures (Da Roit & Weicht, 2013; Van Hooren, 2014).

Research on LIMC arrangements has shown that relatives oftentimes use family terminologies to describe the relationship between the LIMCs and the older care recipient (Anderson, 2000; Ayalon, 2009; Baldassar et al., 2017; Porat & Iecovich, 2010; Teshuva et al., 2019). Treating the LIMC as part of the family has been explained as a strategy to uphold an ideal of care according to which care provided by family carers is the most desirable for the older person in need of care (Pérez-Caramés & Martínez-Buján, 2016; Weicht, 2010). In addition, the use of family terminologies has been identified as a means by which families can dissociate from their role as employers and hence feel less concerned about the LIMCs' working conditions, lack of social security and so on (Tappert & Dobner 2015).

This tendency to present or treat LIMCs as quasi-family carers has been described for countries where families are seen as the main provider of LTC in old age. However, it remains unclear whether it occurs in a similar way in countries where LTC is seen as a public rather than a family responsibility. This chapter addresses this research gap by comparing two countries with very different LTC regimes: Germany and the Netherlands. Compared with Germany, the LTC regime in the Netherlands is less 'familistic', with a broader range of publicly funded care services being offered to dependent older people, and a stronger preference among the population for old-age care to be provided by professional rather than informal caregivers (Maarse & Jeurissen, 2016; Mot et al., 2010).

The different LTC regimes explain why LIMC arrangements are much more prevalent in Germany than in the Netherlands (Böcker et al., 2017; Horn et al., 2019). According to recent estimates, around 230,000 or

M. Bruquetas-Callejo
Center for Migration Law/Sociology of Law, Radboud University Nijmegen, Nijmegen, The Netherlands
e-mail: m.bruquetascallejo@jur.ru.nl

11% of private households with an older person in need of care in Germany are employing LIMCs (Hielscher et al., 2017). By contrast, it is estimated that less than 1000 older persons in need of care in the Netherlands are making use of an LIMC arrangement (Böcker & Bruquetas-Callejo, 2019; Da Roit & van Bochove, 2017). However, the market for this type of LTC service in the Netherlands seems to be growing.

Against this backdrop, the question arises whether these differences are reflected in family carers' expectations of LIMCs and LIMC arrangements. For example, do Dutch family carers, due to the country's highly professionalized LTC regime, expect LIMCs to have similar qualifications as regular home care professionals, whereas German family carers place more importance on qualities that are typically assigned to family carers in which professional knowledge and skills are less relevant than affection, dedication and sacrifice? Another related question is how family carers seek to shape the LIMC arrangement according to these expectations. To explore these research questions, this chapter analyzes data from interviews with family carers in Germany and the Netherlands.

This chapter is structured as follows: In Sect. 2, the German and Dutch LTC regimes are compared to explain the different prevalence and characteristics of LIMC arrangements in the two countries. In Sect. 3, the research methodology is described and in Sect. 4 and 5, the findings from our comparative analysis are presented. In a final section, the findings are discussed in light of the LTC regime approach and conclusions are drawn, including thoughts about future research perspectives.

2 LTC Regimes in Germany and the Netherlands

LTC regime approaches have proved useful to explain the prevalence of migrant care workers in different care sectors across countries (Van Hooren, 2012; Williams & Gavanas, 2008), including the case of LIMCs (Da Roit & Weicht, 2013; Böcker et al., 2017; Horn et al., 2019). The concept of regime refers not only to a set of policies (care, employment etc.) and to their interactions, but also to the relation between policy and culture, and how this relation influences care norms and preferences as

Table 3.1 Key characteristics of the German and Dutch LTC regimes

	Germany	Netherlands
Compulsory LTC insurance	Since 1995	Since 1968
Public expenditure on LTC*	1.5% of GDP	3.7% of GDP
Private expenditure on LTC**	Relatively high (>30% of total spending on LTC services)	Relatively low
Share of LTC recipients aged 65 and over receiving care at home*	74%	66%
Share of people aged 50 and over providing informal care on a daily basis*	8.4%	5.2%

Sources: * OECD, 2019; ** OECD, 2017

well as everyday care practices and experiences (Bettio & Plantenga, 2004; Williams, 2012).

As shown by Table 3.1, LTC regimes in Germany and the Netherlands differ considerably. Although both countries introduced a compulsory LTC insurance scheme, the Netherlands did so nearly 30 years earlier than Germany. Moreover, in the Netherlands a larger share of the spending on LTC services is covered by compulsory insurance and taxes. Public expenditure on LTC as a percentage of GDP in the Netherlands is more than twice the OECD average (1.7%) and nearly two-and-a-half times the percentage of Germany. By contrast, the share of private expenditure (co-payments and out-of-pocket expenses) in total spending on LTC services is relatively high in Germany and relatively low in the Netherlands.

Another difference is that the largest share of public spending (57%) in Germany goes to home-based LTC services, whereas a much larger share of public spending in the Netherlands (75%) is allocated to LTC provided in residential facilities (OECD, 2019). This difference is reflected in a larger proportion of LTC recipients in Germany receiving care at home and in a much larger proportion of LTC recipients in Germany opting for LTC insurance cash benefits rather than in-kind benefits (BMG 2017). In Germany, the option to apply for cash benefits has been offered and promoted from the introduction of the LTC insurance scheme. The use of these benefits is left to the recipients' discretion, and they are often used to pay family caregivers (Jensen & Møberg, 2011). Two-thirds of LTC beneficiaries in Germany, who receive care at home, are cared for solely by close relatives or other informal caregivers (ICF, 2018).

In the Netherlands, cash-for-care benefits were a late addition to the LTC insurance scheme, and LTC beneficiaries still make much more use of in-kind benefits, also for home-based care (MLZ, 2020a, 2020b). The Dutch publicly funded LTC system is more comprehensive than the German system and includes tax-funded social care and home help services for people with lighter care needs. The cash-for-care benefits in the Netherlands are likewise more generous than the ones in Germany, but their use is very tightly regulated and controlled, and they are more often used to buy services from professional care providers (MLZ, 2019).

In brief, policies in Germany place more of the responsibility for the provision of long-term care onto families. This is partly reflected in the rate of people aged 50 years and older frequently providing informal care, which is above the OECD average and considerably higher than the rate in the Netherlands (see Table 3.1). Moreover, there seems to be a correspondence between policies and moral norms in both countries. Surveys found that people in the Netherlands tend to consider the state to be responsible for the provision of care and support for older people, whereas people in Germany tend to believe that the responsibility should be borne equally by the family and the state (Börsch-Supan, 2019). Though a majority in both countries believe that it is preferable for older people to be cared for in their own homes, preferences with regard to who should provide this care differ. In the Netherlands, there is a strong preference for this care to be provided by professional care workers, whereas in Germany, the preference for family care is more pronounced (Eurobarometer, 2007).

Still, over the last two decades, policy reforms in the Netherlands have aimed at substituting publicly funded LTC services by informal care and shifting part of the responsibility for LTC back to families (Grootegoed, Van Barneveld and Duyvendak 2014). Nevertheless, survey results show that in the first decade of the twenty-first century the trend in moral norms was rather toward what Hochschild (1995) called a 'cold-modern' care ideal, according to which working-age woman and men should be in the workforce, with the state bearing the principal responsibility for the provision of care to older people (Van den Broek et al., 2015). This finding suggests a discrepancy with policy goals in the field of LTC. However, it may also be interpreted as being consistent with policy goals in the field of employment. Moreover, the most prevalent old-age care ideal is still a 'warm-modern' one, according to which the family and the state should engage jointly in the provision of LTC (Van den Broek et al., 2015).

In Germany, the norm of family care in old age is nurtured by a moral discourse according to which family care is considered as 'the right thing to do' and by far the better care option. Often based on idealized family relations, family care is socially constructed as 'good care' as it would provide close and affective care relations corresponding to the individual needs and contributing to the well-being of the person in need of care (Horn & Schweppe, 2019). This is surprising in view of the high incidence of violence and neglect in old-age care at home (Eggert et al., 2018; Görgen et al., 2012). Still, families in Germany are faced with a stigmatizing discourse on 'deporting' their frail and elder relatives when moving them to a nursing home (Bender & Schweppe, 2019). In the Netherlands, the image of nursing homes is less negative, making it less of a moral issue for family carers to move relatives with intensive LTC needs to an institutional setting.

The strong reliance on the family as a provider of LTC translates into a relatively higher care burden for family caregivers in Germany (Colombo et al., 2011; OECD, 2017). In a previous study, we found that feeling overburdened was a core reason to hire an LIMC for family carers in Germany. For family carers in the Netherlands who hired an LIMC, a main reason was that the public LTC system could not provide 24/7 care or supervision at home and left them no other option but nursing home care, which they rejected for their relative (Horn et al., 2019).

The different national LTC regimes explain why LIMC arrangements have become widely used in Germany but not in the Netherlands. Moreover, LIMCs in the Netherlands are mostly hired through private intermediary agencies, and due to the high degree of regulation of the Dutch LTC cash benefits, they are seldom employed without a formal contract (Horn et al., 2019). This stands in stark contrast to Germany, where it is estimated that around 90% of LIMCs do not have a regular working contract (Kniejska, 2016). Although the practice of hiring

LIMCs irregularly is widely known, the risk of legal sanctions for families is small as authorities in Germany tacitly tolerate this practice in order to avoid a collapse of the country's LTC regime (Horn and Schweppe in this book; Lutz & Palenga-Möllenbeck, 2010).

However, are these differences between LTC regimes reflected at the micro-level, in family carers' expectations toward LIMCs and in their efforts to shape the care arrangement accordingly? Are German family carers particularly interested in family-like relationships and are they less concerned with complying with, for example, working hours regulations? Do Dutch family carers attach more importance to LIMCs' formal qualifications and are they more concerned about possible breaches of labor law regulations?

3 Data and Methods

This study is part of the comparative research project 'The Emergence and Significance of Transnational LTC Arrangements in Germany and the Netherlands'.[1] To gain insights from various angles, semi-structured interviews with different actors involved in LIMC arrangements (including family carers, care recipients and LIMCs) were conducted between May 2016 and March 2018 in both countries. This chapter is based mainly on the interviews with family carers. They were selected according to their role as primary family carers who organized and managed the LIMC arrangement for the care recipient and, especially in the German case, were often heavily involved themselves in the care for their family member.

For the purpose of this chapter, we analyzed 39 interviews (including follow-up interviews) with 14 primary family carers in Germany and 13 primary family carers in the Netherlands. Follow-up interviews were conducted to clarify questions and capture changes in the LIMC arrangement over time. The same interview guide was used in Germany and the Netherlands. It was structured by three main themes: (1) The reasons for establishing an LIMC arrangement, (2) the everyday provision of care, and (3) the assessment of the quality of the arrangement.

Field entry and sampling was difficult in both countries, although for different reasons. In the Netherlands, the main obstacle was that LIMC arrangements are still rare, so that chain referrals were of little use. Most respondents could only be recruited via LIMC placement agencies, using a clustered sampling technique. In a few cases, the care recipient was already deceased at the time of the interview or the family carer had quit the arrangement for a variety of reasons. In Germany, the main obstacle was the respondents' reluctance to participate because of the irregular employment situation of the LIMCs. Snowball sampling through personal contacts was therefore particularly useful to find interview participants. Attempts to recruit respondents through care associations, parishes, placement agencies and through a Facebook site remained largely unsuccessful. Altogether, in eight care arrangements in Germany the LIMCs were employed irregularly; in five cases, they were placed by an agency; and in one case, the LIMC was directly employed by the family. The different ways to recruit interviewees thus reflect the differences between the two countries concerning the spread of LIMC arrangement and their employment modes as outlined above.

In most cases, the family carers were sons or daughters of the care recipients. In Germany, the children were in their 50s and 60s and varied considerably regarding their socio-economic profiles and family situation. Some lived in the same house as their parent, while others lived several hundred kilometers away. The latter periodically moved into their parents' house to either support or swap with the LIMC. In the Netherlands, the children were of a similar age as their counterparts in Germany, their family situation varied, but they were less heterogeneous in terms of their socio-economic profiles and geographic location. The majority were highly educated, were employed full-time and lived less than 20 km away from the care recipient. In three cases in Germany and two in the Netherlands, the family carer was the care recipient's spouse or partner. All spouses in Germany were males in their 70s and had cared for their wives for several years by themselves before taking the decision to employ an LIMC. The two spouses in the Netherlands were women of working age with a demanding job and/or study.

In some cases, we could also interview the LIMCs. In both countries, all LIMCs came from Central and Eastern European EU member states.

In Germany, the majority came from Poland. In the Netherlands, the majority came from Slovakia. All but one of the LIMCs employed by families in the Netherlands and in Germany were women. Most of them were middle-aged. Some were married; others were divorced or single. Their educational background was very diverse. Only very few LIMCs were trained as nurses or had a professional background in old-age care. Those who were placed by an agency had often received a crash course in caring before taking up their first job.

In most cases, we could not interview the care recipients themselves. Most of them were men or (more often) women in their 70s or 80s. They suffered from conditions such as dementia, Parkinson's disease, old-age frailty or a combination of these. Their care needs varied considerably. Most of them needed continuous supervision and help with activities of daily living. Some of them (also) needed other forms of care.

The data from interviews with family carers were analyzed using qualitative content analysis following Mayring (2010). First, categories were established according to the main research questions: What do family carers expect of LIMCs and through what kind of strategies do they shape the LIMC arrangement? Second, codes were defined to clarify which text components should be included in the analysis. The same scheme for analysis was used to examine the Dutch and German interview material. In a final step, the results from the analyses in both countries were compared to identify differences and similarities.

4 Family Carers' Expectations of LIMCs

The expectations of family carers of LIMCs in both countries show similarities as well as differences. Three aspects seemed to be particularly relevant for family carers' expectations in both countries: their concepts and ideas of 'good care', the care recipients' needs and personality and their own life situations, including their perceived care burden. Above all, family carers expected to establish a well-functioning care arrangement, which enabled the care recipient to stay at home and simultaneously improved their own work-life-family-balance.

When family carers were asked what they found important when employing an LIMC, they recurrently used expressions such as 'the chemistry has to be right' or 'it has to click'. This reflects the importance family carers place on social and relationship aspects and their wish for harmonious and pleasant relationships, in which the ones involved get along well. Expressions such as 'it has to click' also indicate the importance of personality or character traits of the LIMC such as empathy as well as the relevance of shared understandings. Hence, instead of potentially long processes of negotiations to develop desirable relationships, family carers' preferred a 'natural' and thus less complicated match between the LIMC and the care recipient and/or themselves.

With regard to the relationships between the LIMC and the care recipient, the expectations of the family carers differed. Some family carers preferred an emotionally close and affectionate relationship. Others wanted the LIMC to keep a certain distance to the care recipient precisely because of his or her personality and/or difficulties to develop close bonds to an unknown person who lives in the same household. Care recipients were not always willing to accept an LIMC in their household, for example, because they were convinced of still being able to manage activities of daily living by themselves. Especially care recipients who were used to being autonomous and in charge of the household (usually women) were reluctant to pass some if not most of their competencies to the LIMC. In these cases, LIMCs were expected to maintain an appropriate personal distance despite quasi-continuous presence in the household.

Social, personal and relationship aspects were especially pronounced in the interviews with family carers in Germany. Accordingly, they attached rather little or no importance to the LIMC having professional care skills or knowledge. They rather expected LIMCs to have qualities that are often assigned to family carers, such as devotion, a willingness to make sacrifices and affection for the care recipient. Sometimes this included the expectation of LIMCs becoming part of the family and performing tasks beyond the care for the care recipient. An example of this was provided by a family carer (I6, German family carer, husband) who, talking about an LIMC who had 'quickly become part of the family', very much appreciated that she was 'often also taking care of the children [grandchildren]' and participating in family activities (having breakfast together etc.). By contrast, he found it 'pretty extreme' that another LIMC refused to take care of the grandchildren, preferred to eat alone and was waiting for him to come home in the evening so that she could finish her working day. Drawing boundaries and reminding the family carer of the arrangement being an employment relationship with corresponding working hours was thus perceived as an undesirable behavior.

In the Netherlands, too, family carers were inclined to attach foremost importance to LIMCs' social skills. However, Dutch interviewees put somewhat more emphasis on the LIMCs' training and professional care skills. They often complained about the lack of training that agencies provide for care workers. A few family carers hired higher qualified (and more expensive) LIMCs because their relatives needed specialized treatments. Others hired ambulatory home care services to cater for such treatments. Still others simply anticipated that their parents would have to move to a nursing home when their care needs would increase; they saw the LIMC arrangement as a temporary solution. A preference for more professionalized (LIMC) care arrangements emerged in a few interviews, for example, when a family carer explained his ideal as follows: 'I would like my father to have a professional au-pair, paid and supervised by the state' (C14, Dutch family carer, son). Several family carers also reported on discussions they had had with their siblings about whether it was 'responsible' to opt for an LIMC arrangement for their parent and whether a nursing home (with professional care staff) would not be a better option. These examples show that family carers in the Netherlands were more inclined to compare the LIMCs against the standard of trained care workers of home care providers or old-age residential care facilities than their counterparts in Germany.

Family carers in both countries expected LIMCs to manage the household quickly by themselves: 'That's what the person needs to bring. It does not work if someone comes and says: what do I have to do? You can instruct someone and look after her the first week or two but then things have to run by themselves' (I3, German family carer, son). Apart from managing the household, family carers expected LIMCs to provide personal care (e.g., bathing, changing diapers) and company/supervision, and in some cases, they were expected to carry out nursing tasks such as the administration of medication.

Activating the care recipients was also expected from LIMCs by family carers in both countries. For various family carers in the Netherlands, this had been an important motive in opting for an LIMC arrangement instead of an old-age care home. As one respondent put it: ‘These ladies are very good at this. They know how to challenge [people like my father], how to tell them: come on, get moving’ (C1, Dutch family carer, son).

Good language skills were particularly expected by family carers who wished the LIMC to become a conversation partner for the care recipient. This expectation varied according to the care recipient's ability and desire to communicate with other persons on a daily basis. Thus, language skills were seen as less relevant by family carers when the care recipient was not able to engage (or not interested in engaging) in conversations with the LIMC.

Finally, expectations regarding the LIMCs' job content and workload were closely linked to the care recipients' needs as well as the family carers' ability (and willingness) to involve in care and supervision tasks themselves and/or purchase additional care service. Expectations about around-the-clock availability were rather similar among family carers in Germany and the Netherlands, as ensuring continuous supervision for the care recipient was an important reason for them to opt for an LIMC arrangement (Horn et al., 2019).

Whether or not LIMCs will meet the family carers' expectations can hardly be predicted. It remains subject to a process of trial and error. This leads to high turnover rates of LIMCs who either may decide to leave themselves or may be discharged by unsatisfied family carers and/or care recipients. To diminish the risk of frequently changing LIMCs in the household, family carers employ various strategies aimed at establishing a smooth and stable arrangement. The next section describes these strategies.

5 Shaping the Arrangement

An often-found approach to enhance the chances for a smooth and stable LIMC arrangement was that family carers gave the LIMC detailed instructions for how to meet the care recipient's preferences (e.g., regarding cooking, cleaning, hygiene), to deal with his or her peculiarities, to respect his or her space and so on. In the Netherlands, some of these instructions may be included in the 'care plan' that has to be written (by the family carer, sometimes with the assistance of the intermediary agency) and submitted for approval as part of the procedure for obtaining an LTC cash benefit.

Some family carers perceived instructing the LIMC as a time-consuming task, especially when LIMCs changed frequently and/or the family carers were not living close by. Therefore, some opted to write down the instructions. However, as shown by the following quote, such measures were not necessarily successful: 'I made a folder, alphabetically, bathing, what's with the newspaper and so on, so that everyone could look it up. But I have [the feeling] that some cannot read properly or are not able to put it into practice' (I11, German family carer, daughter). A family carer in the Netherlands had a similar experience: 'I taught [the LIMC] precisely how to cook the rice for my mother. But then I saw that she had warmed it up. The rice had hardened, it was not as it should be. (…) Then I got a bit angry' (C13, Dutch family carer, daughter).

Instructing the LIMC has several functions. It certainly serves to make the LIMC familiar with the household, her or his tasks and the routines and habits of the care recipient. It thus helps to avoid ruptures in the care recipient's way of life. Moreover, it lowers the potential for conflicts arising from misunderstandings, failures to take the care recipient's habits into account, violations of boundaries and so on. It can therefore be seen as a preventive measure, which may ultimately pay out for the family carers, who expect to have less problems in the future when giving the arrangement a good start.

However, not all family carers tried to guide the LIMC's performance and her or his relationship with the care recipient through detailed instructions. Some relied on the LIMC's previous experiences in other households. Others simply 'let it flow', as one respondent put it (I1, German family carer, daughter), and only intervened when situations escalated. Such an approach does not necessarily have negative implications for the relationship between LIMC and care recipient. However, it might require a greater effort from LIMCs and care recipients to adapt to each other and increases the likelihood of (unintended) 'mistakes' made by LIMCs. Particularly care recipients who were skeptical about the arrangement may feel confirmed in this by such 'mistakes' and develop a negative attitude toward the LIMC.

Mediating tensions and conflicts between LIMCs and care recipients is another element of family carers' efforts to establish or maintain a well-functioning arrangement. Tensions and conflicts occur for various reasons, for example, when care recipients feel disturbed by the continuous presence of the LIMC, limited in their autonomy and/or patronized by the LIMC. In some cases, this led to verbal aggression and/or humiliations of LIMCs by care recipients. Mediating in such situations can be even more challenging when care recipients are cognitively impaired, so that reasoning with them is difficult. In addition, tensions and conflicts can arise when care recipients dislike or even physically reject tasks performed by LIMCs. In several cases, care recipients refused to be washed or showered by the LIMC. Family carers solved this problem in different ways. They told the LIMC not to insist or to persuade the care recipient, hired an ambulatory home care service for this specific task or took it over themselves.

Approaches to establish and maintain a well-functioning LIMC arrangement also include practices aimed at enhancing the situation of the LIMC. In this regard, family carers used different strategies: they tried to create a pleasant environment and expressed their appreciation and gratitude, verbally as well as through gifts or extra payments. Thus, family carers in both countries emphasized their intention to create favorable working and living conditions to improve the LIMC's motivation and well-being.

How this intention was translated into practice differed between but also within the groups of family carers in the Netherlands and in Germany. Especially family carers in Germany, spouses as well as children, were found to be partly heavily involved themselves in personal hands-on care activities, usually in close cooperation with the LIMC. Their involvement was due to the need for some care activities to be provided by two persons (e.g., lifting the care recipient) and additional care services were not affordable. Relieving the LIMC through their involvement was hence a necessity as well as a strategy to prevent the LIMC from feeling overburdened and exploited.

By contrast, family carers in the Netherlands were less often involved in the provision of hands-on care, which may be attributed to the comprehensive set of publicly funded care services provided by the country's LTC system. In most of the Dutch cases, there was a clear division of tasks between the family carers, who were in charge of organizational and administrative issues, and the LIMCs, who were responsible for the provision of daily care, company and housekeeping. Like their counterparts in Germany, family carers in the Netherlands also coordinated the care network when there were other (professional or informal) care providers involved.

Compared with their counterparts in Germany, family carers in the Netherlands less often thought that the relationships between the LIMC and the care recipient and (especially) between the LIMC and themselves should ideally be 'family-like'. Though some did prefer close and family-like relationships, others preferred the relationships to be friendly, but not too close and informal. An example was a family carer who explained that during his time as an expat in Asia, he had learned that 'if the boundaries [between family and working relationships] are not clear, it is annoying not only for yourself, but also for the person you are paying, with whom you have an employment relationship' (C14, Dutch family carer, son). By contrast, family carers in Germany more often said that LIMCs had become part of the family and they appreciated it when LIMCs used family terminology as well. 'The two younger ones, both are around 40 years old, see my wife as their mother and they say that "this is my family" and "this is my mother" and according to this they care for her' (I2, German family carer, spouse). Nevertheless, a few family carers in Germany emphasized the need not to confound work with family relationships: 'My father said several times, "We manage it as if you [the LIMC] were a member of the family". When I heard this, I thought "Hey stop, it is actually not like that. One has to differentiate a bit. She is not a member of the family. You can treat someone very politely (...) but it remains an employment relationship"' (I10, German family carer, son).

Family carers in the Netherlands were generally more concerned about LIMC working hours and some made specific arrangements to ensure that the LIMC would have a free day. Whereas several family carers in Germany found that it was easy earned money, family carers in the

Netherlands more often saw an incongruence between the work done by the LIMCs and the salaries paid by the intermediary agencies: 'For such a [low] wage, I consider it a whole lot of work' (C1, Dutch family carer, son). Several family carers in the Netherlands negotiated with the agency to obtain a pay raise for the LIMC, while others moved to agencies that offered better wages for LIMCs. The LIMCs' salaries also played a role in Germany, but in a different way. Although most family carers were well aware of the LIMCs' sacrifice and precarious situation, they did not utter that the pay was inappropriate. This difference might indicate that family carers in the Netherlands are more inclined to compare LIMCs to regular home care professionals, while in Germany LIMC arrangements are rather seen as similar to family care, with the downplay of burdensome and straining demands and challenges that comes with it.

6 Discussion and Conclusions

Family carers in Germany and Netherlands expect LIMCs to adapt to the care recipients' personality and to offer him or her as much continuity as possible with regard to his or her habits and routines. Given the individuality of every care recipient, these widespread expectations can entail sets of very specific expectations, which vary from case to case and which may also change with the care recipient's care needs and cognitive health.

Differences in expectations between family carers in Germany and the Netherlands were found regarding the professional training of LIMCs and the desired type of relationship between the LIMC and the care recipient and her or his family, respectively. Family carers in the Netherlands tended to see old-age care more as a profession whereas family carers in Germany rather associated it with qualities that are typically assigned to family care, such as devotion and sacrifice. This difference may be explained by the recognition (and degree of professionalization) of care work and the distribution of responsibilities in the Dutch and German LTC regime. Compared with the Netherlands, where LTC is seen as a public responsibility, to be delivered by professional care workers, in Germany it is primarily seen as a private issue, with LTC being overwhelmingly provided by family caregivers. This implicit analogy between LIMCs and informal caregivers may nurture expectations that LIMCs should perform their tasks as if they were family carers (e.g., being constantly available).

Related to this, family carers in Germany were indeed more likely to appreciate the development of family-like relationships while family carers in the Netherlands more often showed a preference for friendly but not too close and informal relationships. However, family-like relationships are neither generally expected by family carers in Germany nor do they develop quasi-automatically over time due to the specificities of the arrangement (living under the same roof, sharing daily routines etc.). This relativizes findings from studies, which suggest that family terminology and kinning processes are constitutive elements of LIMC arrangements (Anderson, 2000; Baldassar et al., 2017) and strengthens findings from previous studies on Eastern European LIMCs in Germany (Karayali, 2010). It indicates the necessity to carefully consider the complex constellation of factors influencing the development of relationships in LIMC arrangements. As shown by our study, care recipients may be reluctant or unable to establish closer bonds with LIMCs, for example, because they reject the care arrangement or because of language barriers (cf. Teshuva et al., 2019; Timonen & Doyle, 2010). Similarly, LIMCs may not be interested in becoming part of the family and may prefer not to blur the boundaries between formal paid and informal (and additional) unpaid work.

Moreover, family carers in both countries have been shown to use various approaches and strategies to enhance the chances that their expectations of the LIMC and the care arrangement are fulfilled. Detailed instructions are given to the LIMC to make the transition as smooth as possible. Family carers also develop strategies to prevent conflicts which could burden the LIMC's relationship with the care recipient and might require their mediation or even mean having to search for a new LIMC.

Still, how family carers (attempt to) shape the arrangement does not only depend on their personal abilities to prevent and solve conflicts and build harmonious relationships. It is also structured by the surrounding LTC infrastructures and the families' economic means, which may or may not enable them to make use of additional LTC and home help services. Thus, the LTC cash benefits in the Netherlands, which are more tightly regulated but also more generous than the ones in Germany, enable family carers in the Netherlands to hire LIMCs through relatively expensive private placement agencies. These agencies may manage family carers' expectations, for example, by monitoring the working conditions of the LIMCs and intervening when they risk to become overburdened. More generous cash benefits may also make it possible to incorporate the LIMC into a broader network of professional care and home help services. By contrast, the German LTC system provides families with only limited access to affordable services to complement the LIMC arrangement. Consequently, only economically better-off families in Germany can reduce the LIMCs' (and their own) workload through making use of additional old-age care services.

To conclude, differences in the German and Dutch LTC regime seem to be partly reflected in family carers' expectations of LIMC arrangements and their efforts to shape these arrangements. However, rather than clear general patterns, our findings indicate nuanced differences, which deserve more detailed exploration in future research. Aspects requiring further investigation include the association between modes of employment (regular vs. irregular and with or without intermediaries acting as employers for the LIMCs) and the LIMCs' working conditions and type of relationship with care recipients and family carers, respectively. What our study reveals is the complex entanglement of factors, which have a bearing on family carers' expectations of and approaches to shaping LIMC arrangements, including their perception of what is the best possible arrangement for their dependent older relative as well as for themselves. These perceptions are highly individual but at the same time dialectically interwoven with social norms about old-age care and with the availability of affordable LTC and home help services.

Acknowledgments The research project 'Emergence and Significance of Transnational Elderly Care Arrangements' (ESTRANCA) has been enabled by research funding from the German Research Foundation (DFG) and the Netherlands Organisation for Scientific Research (NWO) in the framework of the Open Research Area (ORA) for the social sciences.

Note

1. For the Dutch part, the study was reviewed and approved by the Ethics Committee of the Faculty of Law and Nijmegen School of Management of Radboud University Nijmegen (registration number 2016.11). For the German part, no ethical approval was requested by the Johannes Gutenberg University of Mainz.

References

Anderson, B. (2000). *Doing the dirty work? The global politics of domestic labor*. Zed Books.

Ayalon, L. (2009). Family and family-like interactions in households with round-the-clock paid foreign carers in Israel. *Ageing & Society, 29*, 671–686.

Baldassar, L., Ferrero, L., & Portis, L. (2017). 'More like a daughter than an employee': The kinning process between migrant care workers, elderly care receivers and their extended families. *Identities, 24*(5), 524–541.

Bender, D., & Schweppe, C. (2019). Care facilities for Germans in Thailand and Poland: Making old age care abroad legitimate. *International Journal of Ageing and Later Life, 13*(2), 115–143.

Bettio, F., & Plantenga, J. (2004). Comparing care regimes in Europe. *Feminist Economist, 10*(1), 85–115.

BMG. (2017). Zahlen und Fakten zur Pflegeversicherung. Stand 20 October 2017. www.bundesgesundheitsministerium.de/fileadmin/Dateien/3_Downloads/Statistiken/Pflegeversicherung/Zahlen_und_Fakten/Zahlen_und_Fakten.pdf.

Böcker, A., Horn, V., & Schweppe, C. (2017). Old age care regimes and the emergence of transnational long-term care arrangements for the elderly. In L. Good Gingrich & S. Köngeter (Eds.), *Transnational social policy, social welfare in a world on the move* (pp. 222–242). Routledge.

Böcker, A., & Bruquetas-Callejo, M. (2019). Inwonende buitenlandse zorgverleners. Opkomst en betekenis van een nieuw type zorgarrangement in de Nederlandse ouderenzorg. Nijmegen sociology of Law Working Papers 2019/02.

Börsch-Supan, A. (2019). Survey of Health, Ageing and Retirement in Europe (SHARE) Wave 2. Release version: 7.1.0. SHARE-ERIC. Data set. DOI: https://doi.org/10.6103/SHARE.w2.710

Colombo, F., Llena-Nozal, A., Mercier, J., & Tjadens, F. (2011). *Help wanted? Providing and paying for long-term care*. OECD.

Da Roit, B., & Weicht, B. (2013). Migrant care work and care, migration and employment regimes: A fuzzy-set analysis. *Journal of European Social Policy, 23*(5), 469–486.

Da Roit, B., & van Bochove, M. (2017). Migrant care work going Dutch? The emergence of a live-in migrant care market and the restructuring of the Dutch long-term care system. *Social Policy & Administration, 51*(1), 76–94.

Eggert, S., Schnapp, P., & Sulmann, D. (2018). Aggression und Gewalt in der informellen Pflege. Retrieved July 25, 2020, from https://www.zqp.de/wp-content/uploads/ZQP_Analyse_Gewalt_informelle_Pflege.pdf

Eurobarometer. (2007). *Health and long-term care in the European Union (Special Eurobarometer 283)*. European Commission.

Görgen, T., Herbst, S., Kotlenga, S., Nägele, B., & Rabold, S. (2012). *Kriminalitäts- und Gewalterfahrungen im Leben älterer Menschen. Zusammenfassung wesentlicher Ergebnisse einer Studie zu Gefährdungen älterer und pflegebedürftiger Menschen*. Bundesministerium für Familie, Senioren, Frauen und Jugend.

Grootegoed, E., Van Barneveld, E. & Duyvendak, J.W. (2014). What is customary about customary care? How Dutch welfare policy defines what citizens have to consider 'normal' care at home. *Critical Social Policy, 35*(1), 110–131.

Hielscher, V., Kirchen-Peters, S., & Nock, L. (2017). *Pflege in den eigenen vier Wänden: Zeitaufwand und Kosten*. Hans-Böckler-Stiftung.

Hochschild, A. R. (1995). The culture of politics: Traditional, postmodern, cold-modern and warm-modern ideals of care. *Social Politics, 2*(3), 331–346.

Horn, V., Schweppe, C., Böcker, A., & Bruquetas Callejo, M. (2019). Live-in migrant care worker arrangements in Germany and the Netherlands: Motivations and justifications of family decision-making. *International Journal of Ageing and Later Life, 13*(2), 83–114.

Horn, V., & Schweppe, C. (2019). Gewaltpotentiale und Gewaltschutz in Privathaushalten mit 24-Stunden-Pflegekräften. *Zeitschrift für Sozialpädagogik, 18*(4), 394–413.

ICF. (2018). Peer review on 'Germany's latest reforms of the long-term care system'. Synthesis Report. Brussels: European Commission, Directorate-General for Employment, Social Affairs and Inclusion. https://ec.europa.eu/social/main.jsp?langId=en&catId=89&newsId=9008.

Jensen, P. H., & Møberg, R. J. (2011). Tensions related to the transition of elderly care from an unpaid to a paid activity. In B. Pfau-Effinger &

T. Rostgaard (Eds.), *Care between work and welfare in European societies. Work and welfare in Europe* (pp. 98–114). Palgrave Macmillan.

Karayali, J. (2010). *Transnational haushalten. Biografische Interviews mit care workers aus Osteuropa*. VS Verlag für Sozialwissenschaften.

Kniejska, P. (2016). *Migrant Care Workers aus Polen in der häuslichen Pflege: Zwischen familiärer Nähe und beruflicher Distanz*. VS Verlag für Sozialwissenschaften.

Lutz, H., & Palenga-Möllenbeck, E. (2010). Care work migration in Germany: Semi-compliance and complicity. *Social Policy & Society, 9*(3), 419–430.

Maarse, H., & Jeurissen, P. (2016). The policy and politics of the 2015 long-term care reform in the Netherlands. *Health Policy, 120*(3), 241–245.

Mayring, P. (2010). *Qualitative Inhaltsanalyse* (12th ed.). Beltz Verlag.

MLZ. (2019). Kerncijfers Wlz-pgb naar soort overeenkomst, 2015 t/m 2017. www.monitorlangdurigezorg.nl/publicaties/maatwerk-publicaties/2019/08/01/kerncijfers-wlz-pgb-naar-soort-overeenkomst-2015-t-m-2017.

MLZ. (2020a). Kerncijfers. Gebruik Wlz-zorg in natura. www.monitorlangdurigezorg.nl/kerncijfers/gebruik/gebruik-wlz-zorg-in-natura.

MLZ. (2020b). Kerncijfers. Gebruik persoonsgebonden budget. www.monitorlangdurigezorg.nl/kerncijfers/gebruik/gebruik-persoonsgebonden-budget.

Mot, E., Aouragh, A., De Groot, M., & Mannaerts, H. (2010). The Long-Term Care System for the Elderly in the Netherlands. ENEPRI Research. Report No. 90. Brussels: Centre for European Policy Studies.

OECD. (2017). *Health at a Glance 2017: OECD Indicators*. OECD Publishing.

OECD. (2019). *Health at a Glance 2019: OECD Indicators*. OECD Publishing.

Pérez-Caramés, A., & Martínez-Buján, R. (2016). Migrant care and the production of fictive kin. In C. Kroløkke, L. Myoing, S. Willum Adrian, & T. Tjørnhøj-Thomsen (Eds.), *Critical kinship studies* (pp. 83–100). Rowman & Littlefield.

Porat, I., & Iecovich, E. (2010). Relationships between elderly care recipients and their migrant live-in home care workers in Israel. *Home Health Care Services Quarterly, 29*(1), 1–21.

Tappert S., & Dobner M. (2015). 'Being a Member of the Family?' Meanings and Implications in Paid Migrant Domestic and Care Work in Madrid. In: M. Kontos & G.T. Bonifacio (Eds.), *Migrant Domestic Workers and Family Life* (pp. 276–299). London: Palgrave Macmillan.

Teshuva, K., Cohen-Mansfield, J., Iecovic, E., & Golander, H. (2019). Like one of the family? Understanding relationships between migrant live-in care workers and older care recipients in Israel. *Ageing & Society, 39*(7), 1387–1408.

Timonen, V., & Doyle, M. (2010). Migrant care workers' relationships with care recipients, colleagues and employers. *European Journal of Women's Studies, 17*(1), 25–41.
Van den Broek, T., Dykstra, P. A., & Van der Veen, R. J. (2015). Care ideals in the Netherlands: Shifts between 2002 and 2011. *Canadian Journal on Aging, 34*(3), 268–281.
Van Hooren, F. (2012). Varieties of migrant care work: Comparing patterns of migrant labour in social care. *Journal of European Social Policy, 22*(2), 133–147.
Van Hooren, F. (2014). Migrant care work in Europe: Variety and institutional determinants. In M. Léon (Ed.), *The transformation of care in European societies* (pp. 62–82). Palgrave Macmillan.
Weicht, B. (2010). Embodying the ideal carer: The Austrian discourse on migrant carers. *International Journal of Ageing and Later Life, 5*(2), 17–52.
Williams, F., & Gavanas, A. (2008). The intersection of childcare regimes and migration regimes: A three-country study. In H. Lutz (Ed.), *Migration and domestic work: A European perspective on a global theme* (pp. 13–28). Ashgate.
Williams, F. (2012). Converging variations in migrant care work in Europe. *Journal of European Social Policy, 22*(4), 363–376.

4

The 'Care Mix' Adopted by Belgian Families and the Growing Presence of Migrant Workers in Old Age Care in Belgium

Chiara Giordano

1 Introduction

The economic, demographic and societal transformations that occurred in Europe in the last few decades—such as the aging of the population, the increased female participation rates in the labor market, changes in family structures and the increased mobility of family members, among others—have profoundly affected not only the demand for care for older people, but also the organization of care provision (Bettio et al., 2006; Gerhard et al., 2005; Simonazzi, 2009; Anderson, 2012; Ambrosini, 2013; Anderson & Shutes, 2014). Furthermore, as scholars have emphasized, the reinforcement of neoliberal ideas and the retrenchment of welfare states, as well as the constraints imposed by the economic crisis, have influenced the recent reforms in old age care in all European countries (Guo & Willner, 2017; Schwiter et al., 2018).

C. Giordano (✉)
Group for Research on Ethnic Relations, Migration and Equality, University of Brussels, Brussels, Belgium
e-mail: chiara.giordano@ulb.ac.be

V. Horn et al. (eds.), *The Global Old Age Care Industry*,
https://doi.org/10.1007/978-981-16-2237-3_4

In Belgium, the 6th State Reform in 2014, which corresponds to the last institutional reform in the field of welfare, brought about a transfer of responsibilities and a reorganization of services for older people and contributed to shaping the contemporary old age care system. While these institutional reforms reinforced some trends that were already visible starting from the 1980s, they also aligned the Belgian old age care system with trends that are common to other EU countries. Among these trends are an overall shift of responsibilities from the national to the local level, the strengthening of home care services, and the privatization of services (Rostgaard, 2007; Pavolini & Ranci, 2008; Rummery, 2009). The most visible consequences of these trends have been the growing complexity of care provision and the multiplication of actors in the old age care sector.

Today, as in other Western countries, the system of care provision for older people in Belgium involves an increasing number of actors—formal, informal, public and private ones—each of them providing different types of services, at different prices, and addressing various needs. The growing care and domestic needs of families, together with the growing complexity in the provision of services, have put a strain on families, who are more and more encouraged to find alternative solutions. As observed in other countries, this typically results in the adoption of an individual 'care mix' that combines the various options at their disposal (Naldini et al., 2013).

Despite the general lack of acknowledgment of this phenomenon, both in the literature and at policy level, the increasing ethnicization of this sector is more and more visible, especially in the Region of Brussels (SPF Emploi & UNIA, 2017). Female migrants, who represent a cheaper and more flexible labor force, compared to natives, are increasingly employed either by formal (public and private) care providers or directly by families.

This chapter explores the main trends in the field of old age care in Belgium and the consequences that these trends have on families' choices, on the one hand, and on the recruitment of migrant caregivers in the Region of Brussels, on the other hand. The discussion is based on the analysis of in-depth interviews with key informants working in the field of old age care. Three main features of contemporary old age care in

Belgium are discussed at length: (1) the strengthening of home care servicesc, (2) the privatization of services (both residential and home-based) and (3) the fragmentation and multiplication of care providers active in the domiciliary old age care sector in Brussels. Starting from the narratives of the respondents, I discuss some of the consequences that these three trends have on the actual choices operated by care receivers and their families in Brussels, as well as on female migrants working in this sector.

2 Old Age Care in Belgium

The Belgian welfare system is one of the most service-oriented in Europe and is generally believed to adequately cover the care needs of families (European Commission, 2007). The state intervention in the field of old age care is traditionally important, both in terms of provision of services and in terms of public spending (Willemé, 2010; Degavre & Nyssens, 2012). According to the OECD data, the Belgian public expenditure on long-term care (LTC) represented about 2.2% of the total GDP in 2017, a figure among the highest in Europe, just behind Northern countries.

To use the well-known classification of Esping-Andersen (1990), the Belgian welfare regime includes features of both the social-democratic welfare regimes and conservative corporatist welfare regimes[1] (Palier, 2006; Gilain & Nyssens, 2001; Willemé, 2010; Degavre & Nyssens, 2012). Among all types of old age care provision, residential services represent the core of the public old age care provision in Belgium, while home care services and financial allocations have been traditionally more marginal. Indeed, as observed by scholars, the Belgian old age care system is traditionally based on a strong interconnection between state intervention and the fundamental role of family care providers (Leitner, 2003; Degavre & Nyssens, 2012). In this sense, it can be considered a mixed system, in that it combines extensive publicly financed formal services with significant informal care (Willemé, 2010).

Starting from the 1980s and increasingly in recent years, new configurations started to appear in the way residential and home-based old age care are organized and in terms of the actors involved in the provision of care. In Belgium, the institutional and administrative changes that occurred in recent years, and in particular those linked to the 6th State Reform in 2014, fostered a reorganization of the old age care system. Among the most visible trends, three can be identified as having an impact on the provision of care services, on families and workers alike.

The first is the shift of responsibilities from the national to the local level (Rummery, 2009; Ranci & Pavolini, 2015), which in Belgium resulted in a transfer of responsibilities from the federal State to the Communities and Regions in the 1980s and in the regionalization of all responsibilities in the field of home care in 2014 (Dumont, 2017). This shift, which was accompanied by a transfer of responsibilities between sectors (specifically, from the health to the social sector), had repercussions in the provision of old age care services. Although the consequences are still difficult to analyze at present, the major risk is the development of a model of care provision specific to each region (Brussels, Wallonia and Flanders), each offering different types of provision, entitlements, insurance and financial allocations, depending on the available resources and the political will of each region.

Second, in parallel with the general trend toward the 'de-institutionalization' in the field of health care (Willemé, 2010), the emphasis started to be placed on domiciliary care, both in the public discourse and in the policy arena. Even if residential services remained the core of the Belgian old age care system, existing home care services were reinforced and new (public and private) initiatives were established. As in other European countries, the main reasons behind this shift were to reduce the public expenditure linked to residential care, on the one hand, and to offer older people the possibility to exercise their 'freedom of choice', on the other hand (Gilain & Nyssens, 2001; Willemé, 2010).

The third trend is the increasing privatization of both residential and home care services, which started to become visible from the 1990s. In Belgium, the privatization of care services is realized in two ways: firstly, through the public subsidization of both non-profit and for-profit organizations, through a system of accreditations, and secondly, through the establishment of purely private initiatives, which are not subsidized by the state and not bound to state regulation. While the first type of privatization concerns both residential and home care services, the second type of privatization only concerns the home care sector.[2] Given this complex system of subsidization and the establishment of an increasing number of private initiatives, today the actors intervening in the field of old age care are extremely differentiated. While the residential sector is quite homogeneous, the greatest variation is observed in the home care sector, where the range of services available to families include public, semi-public, private, formal and informal services.

3 Formal Old Age Home Care Services in the Region of Brussels

Looking specifically at formal home-based care,[3] the main difference between care providers is determined by whether they have an accreditation with public bodies. For a purpose of simplification, in this chapter I will refer to non-profit organizations that have an accreditation with a public body as 'public' care providers and to all for-profit or non-profit organizations without a public accreditation as 'private' providers.

Public care providers, which receive a public funding in the form of a fixed number of subsidized hours per year, are required to comply to a common state regulation that determines, among other things, the number of hours and the type of service to which beneficiaries are entitled, the duration and the timetable of services (usually limited to day-time, excluding nights and weekends), the activities authorized, the profiles of care workers, their level of qualification, as well as their employment conditions. These non-profit organizations offer prices based on the beneficiary's income, according to institutional scales, and represent the less expensive option available to families.[4] The professional profile of workers, as fixed by the public regulation, includes health assistants (*aides-soignantes*), family assistants (*aides familiales*), housework attendants (*aide-ménagères sociales*) and sometimes nurses.

The care providers that do not have any accreditation with public bodies include both non-profit and for-profit organizations and represent a more heterogeneous group. Since these care providers are not subjected to institutional regulations, they respond to different needs and different beneficiaries and offer services that vary enormously in terms of price and type of services. Since the profile of workers is not fixed by law, they usually work under the common label of care/health attendants (*garde-malade*).

Despite the diversity of these initiatives, three types of private providers appear to be increasingly present in the Region of Brussels. The first includes private agencies offering general old age care services and working with a pool of self-employed workers. The second includes online platforms working under the legal framework of the 'collaborative economy', which allows private persons to offer care services (without any employment contract), in exchange of a non-imposable payment, up to a certain amount per year. The third includes private organizations, with their headquarters inside or outside the national territory, offering live-in 24-hour services to older people and recruiting female caregivers in other European countries, particularly in Eastern European countries.

Finally, the private agencies working under the voucher system (*titres-services*) represent a further initiative, separate from those mentioned above, and constitute one of the alternatives most known and used by families. This system, which was implemented in 2004, allows Belgian families to buy vouchers—each of them corresponding to one hour of cleaning services—at a cost of around 9–10 euros per hour. The significant subsidization by the state, together with the tax deduction to which families are entitled, makes the vouchers a relatively cheap option. While the state limits the number of vouchers to a maximum of 500 per person per year, the number of vouchers allowed for older people is extended to 2000 hours per year. Although the services offered under this system only include housework services—and strictly exclude care services—vouchers are increasingly used to cover care activities (Willemé, 2010).

A common feature shared by all care providers in the Region of Brussels—be them for-profit, not-for-profit, with or without a state subsidy—is that most caregivers have a migrant background. Because official data on the country of birth and the origin of workers are only available for workers under the voucher system[5] (Idea Consult, 2018) and data on the origin of workers are not collected systematically by single care service providers, it is not possible to quantify the presence of migrants and/or second generations in the workforce.[6] However, according to the participants in this study the overrepresentation of caregivers with a migrant background is increasingly visible both in the public and in the private sector, to the point that in certain organizations they represent the entire workforce.

It is important to notice that the information provided by the respondents with this respect is often based on their 'perception' of the migrant background of caregivers (e.g. based on the name, accent or the physical traits of caregivers) rather than their actual demographic data. According to this qualitative assessment, the ethnicization of the workforce is equally visible in the public and private sector, but differs with respect to the country of origin of caregivers, as well as the duration of caregivers' settlement in Belgium. In the public sector (both residential and home-based), the majority of caregivers are of African origin and include both foreign-born and second-generation women. In the private sector, the profile of migrant caregivers is much more differentiated and includes a larger share of foreign-born women, often from Eastern European countries and Latin America, but also from North Africa, Sub-Saharan Africa and continental Europe.

4 Methodology

The empirical material used for this chapter is part of a wider research on old age care in Brussels, carried out with quantitative and qualitative methods, which includes interviews with migrant caregivers, older people (and families) and care service providers. The material used for this chapter includes 15 in-depth interviews with key informants in the formal old age care sector in Brussels, conducted in the period between October 2018 and May 2019.

The choice of the participants was based on multiple criteria. The first was to cover the diversity of care providers in the Region of Brussels, as well as the most used and well-known services, based on a preliminary analysis of the literature on Belgian care services and on the available information at the institutional level (texts of legislation, online web sites of institutional actors, etc.). The second was to include both public and private care providers, as well as respondents from different political orientations. The third criterion was to include both respondents working in the entire territory of the Brussels' Region, and respondents working at a more local level (i.e. the different municipalities). This is due to the fact that characteristics of certain neighborhoods—and especially the socio-economic status and the origin of the population—may influence not only the type of beneficiaries and/or workers and the demand for old age care, but also the provision of services.

The respondents include the Federation of home care services in Brussels (FSB); a trade union *Fédération Générale du Travail de Belgique* (FGTB); a non-profit organization, providing counseling on residential care for older people (*InforHomes*); the Brussels' Regional Ministry of Economy and Employment (*Bruxelles Economie et Emploi*), responsible for the vouchers system in Brussels; the service in charge of home care services of the *Commission Communautaire Française* (COCOF); four private care providers; and five public care providers, of which two are active in the 19 municipalities of Brussels and three only at the municipality level (Woluwé-Saint-Pierre, Anderlecht and Saint-Gilles).

The choice of the three municipalities was based on the analysis of the socio-demographic characteristics of the 19 municipalities of Brussels (Statbel, 2017). The municipality of Woluwé-Saint-Pierre includes mainly residential neighborhoods and is characterized by a high concentration of older people, with a high socio-economic level and of Belgian origin. The municipality of Anderlecht includes both low- and middle-class neighborhoods, the latter characterized by a strong concentration of older people, mainly of Belgian origin. The municipality of Saint-Gilles includes a younger population with a low socio-economic level, mainly of foreign origin.

All interviews were organized by email or telephone and were realized in the working environment of the respondents. Most of the interviews were conducted with only one respondent—in most cases the director or the person in charge for the service—and only two interviews involved more than one person. Each interview, lasting between one and half hour to three hours, was then entirely transcribed (verbatim) and analyzed with 'thematic analysis', as described by Ramos (2015). The extracts reported in this chapter have been translated by the author from French to English.

5 The Reinforcement of Home Care Services

In line with what is known from the literature on care regimes, the professionals interviewed in the framework of this study overall confirm that the strengthening of old age home care services responds both to the wish of care receivers to remain in their homes as long as possible and to the economic imperative of reducing public costs (Rostgaard, 2007). In both cases, this is found to have direct consequences on families' options and choices and on the recruitment of migrant caregivers.

While all respondents share the same belief that home-based care responds to the actual wish of older people to remain at home, some of them point at the negative consequences that this 'rhetoric' may have on families and on workers alike.

> *I believe there is nevertheless a sort of demagogic aspect there—and I dare to say it—because many policies started from the assumption that people want first and foremost to stay at home. [...] To stay home at all costs can be extremely harmful to everyone, starting from the older person herself. [...] For someone who is highly dependent, who really needs 24-hour care, with regular visits during the day, but also at night, it must be clear that home care is not necessarily covered, since it is difficult to find night staff, and when you find this 24-hour home care services cost more than the rest homes. [...] It's a very dangerous rhetoric. It makes you feel guilty. [...] It's an argument that also reinforces the guilt of families: that's terrible!*

On the one hand, the emphasis on home care services as the perfect solution can have economic and social disadvantages for families. Not only domiciliary care can be considerably more expensive for families, especially when the degree of dependency requires full-time assistance, but families may be faced with social stigma when they cannot allocate enough time and/or resources for the care of their older members at home. Social pressure, together with other factors, can encourage them to adopt informal solutions, such as the direct employment of (often migrant) older carers, which is seen as a more suitable option than residential care. Therefore, similar to Mediterranean countries, the emphasis placed on home care can potentially generate—or reinforce—the social aversion for residential old age care and the preference for in-house informal employment, a solution considered more acceptable from the point of view of intergenerational obligations (Catanzaro & Colombo, 2009; Ambrosini, 2013).

On the other hand, to meet the wish of older people to remain at home even when health conditions require full-time assistance means that the amount of care required cannot be covered by public services. Because of their vulnerability in the labor market, among other factors, migrants are more likely than locals to be employed as night and weekend staff, and even more so when 24-hour shifts and live-in work are required. In this sense, the strengthening of home-based services ultimately encourages families either to opt for private providers, which offer longer shifts, or to directly employ informal caregivers. In both cases workers are more likely to be female migrants.

Concerning the economic argument, the reorganization of old age care services is considered as part of the overall imperative of reducing public costs, in line with the general shift toward neoliberal economic measures in all Western societies (Schwiter et al., 2018). Although in Belgium austerity measures are less visible than in Southern Europe, most respondents argue that the shift from residential to home care services is essentially due to the will to reduce costs. According to some of the respondents, this shift is one of the consequences of the neoliberal measures that have largely affected social protection and that are particularly visible in the field of health care (i.e. reduction in the number of beds in hospitals, reorganization of health care structures along NPM lines, etc.).

However, the idea that old age home care implies a reduction of public costs, compared to residential care, is not shared by all respondents. On the contrary, the cost for the state depends on the intensity of the use of home care services: if costs can be reduced when the older person only requires a limited number of hours, a more intense use of services can be considerably more expensive for the state. Therefore, it is rather the introduction of regulations limiting the entitlements to public home care services—a phenomenon which already started in the 1980s (Gilain & Nyssens, 2001)—that constitutes the actual strategy for reducing public expenditure for old age care. From the point of view of households, the major consequence is the reduction in the number of hours entitled to each beneficiary. The result is that in the case of a strong degree of dependency the 'free choice' of older people to stay home ultimately falls on families' shoulders, as only a basic number of hours is covered by public providers, while the majority has to be filled with alternative options, either private services or family help.

It is interesting to notice that the impact of the economic argument on the recruitment of migrants acts on two levels simultaneously. First, the limitation in the entitlements of public home care encourages the recruitment of migrant caregivers, either through private agencies offering long services, up to 24 hours, or through the direct employment of informal caregivers, depending on the resources of families. Second, the reduction of costs also affects the working conditions of all caregivers, whose wages remain highly uncompetitive compared to other jobs. This applies to both public and private providers and can help explaining the presence in this sector of the most vulnerable segments of the labor market, which in Brussels correspond to migrants, regardless of their level of education.

6 The Privatization of Services

To explore the consequences of the privatization of old age home care services on families and on workers, I start from the arguments used by the respondents to highlight the advantages and the disadvantages of public and private services respectively.

The main argument in favor of privatization is linked to the adaptability and the flexibility of private services, which are conceived as to be entirely tailored to the needs of the older person, without the restrictions imposed by state regulations. This flexibility covers many aspects linked to the care of the older person: the type of services provided, the length and intensity of services, as well as their timetable, which typically fills the gaps left by public providers. Contrary to public care providers, who make a clear distinction between what can and cannot be performed by each worker, private providers offer custom-made services, where the same worker can cover a wide range of needs of the care recipient, ranging from keeping company, housework, hygiene tasks, up to semi-medical acts. Also, services do not have restrictions in terms of time and typically cover nights and weekends, up to 24-hour care, if required. As explicitly stated by private providers, the lack of regulations is considered a clear competitive advantage.

> *Here we can really do whatever we want [...] because there are no rules. Actually it's not—how could I say?—it's not recognised. So, this is an advantage compared to subsidised things.* (Private provider)

> *Of course, it is an asset to be able to be more flexible with this respect, to keep some freedom ... this is really good.* (Private provider)

> *They [the workers] can actually do everything ... everything that a member of the family would do, without asking themselves: "Am I allowed or not?"* (Private provider)

The main consequence is that private providers tend to promote a vision of home care work as a sort of continuation of family help, where the worker can cover the entire needs of the care recipients, just as a family member would do. Instead of questioning the traditional distinction between formal/informal, paid/unpaid, public/private (Anderson, 2000; Lutz, 2008), private care providers build their services precisely on the ambiguous position of home care.

However, the privatization of home care services is associated to a series of negative consequences, which are highlighted by public care providers and by the other interviewed experts (trade unionists, representatives of governmental bodies, etc.). First, if private services may be more flexible in meeting the needs of the beneficiaries, the lack of regulations translates into a lower quality of services and an increased risk for both caregivers and care receivers. Second, the economic logic that guides for-profit companies is not compatible with care work. The following extract clearly expresses the ethical issues linked to the commodification of old age care and the risks associated to it.

> *This phenomenon of commodification has a crucial impact for us. [...] This commodification of the offer, I think, is very problematic from an ethical point*

of view, because actually it makes the older person in a role of customer. But he is not a normal customer. So, it's true that if we go to buy a mobile phone somewhere and we are not happy with it, it doesn't matter, we go elsewhere. Here we cannot make a mistake, it is not a consumption good. [...] And we also realise that at the end of the day everybody is suffering.

Regardless of whether the differences between public and private providers concern ethical questions or the type of service provided (quality, content, duration, etc.), privatization seems to have specific consequences for both households and workers. While the main risks for families are linked to the affordability and the quality of services, the consequences for caregivers affect at least two aspects. First, working conditions in the private sector are considerably poorer than those in the public sector, with all respects: wages, time schedules, employment benefits and job security, among others. Second, caregivers employed by private care providers are more likely to be victims of discriminations. While public care providers work on a system of rotation of care workers—a system which is both internally constrained and voluntarily chosen—and families have no say in the choice of the workers, private providers accord families the possibility to express their preferences and to 'choose' the most suitable worker. This has dramatic consequences in terms of gender and ethnic discriminations.

The caregivers employed in private organizations appear to be the ones who suffer from the most explicit (and sometimes violent) forms of discriminations, which are mainly directed toward non-Europeans, people of color and Muslims. Privatization can thus initiate a vicious circle: while migrants—and among them the most vulnerable—are increasingly attracted to private providers, because often they do not possess the required qualifications and they are more likely to accept unsuitable hours and low wages, it is precisely in private companies that they are more likely to be discriminated against. While racism is reported as being a behavior specific to care recipients and families, employers themselves may reproduce some forms of discrimination. This is achieved either by justifying and indulging the racist requests of beneficiaries, in the name of their mental or health conditions, or by constructing a hierarchy of desirability of workers, just as families do.

> *When they [the caregivers] go to someone with Alzheimer, when I send an African lady, she tells me: "Look, he called me nigger, he said that we are stealing their jobs". But these people have Alzheimer, they don't know what they are saying, they just repeat things they've learned.* (Private provider)

> *Let's take the example of Poland, for instance. Often Polish women speak good French, but the problem of Polish people is that it is never for long periods. […] Maybe it's linked to the culture, which is more fearful [Polish people are more suspicious]. … Romanians are people that … well, you need to get their trust, at the beginning they are very suspicious, because I think they have been fooled often. So, you really need to—as I always say—to tame them. […] We also worked a little with Portugal. The problem—and it is very clear—is that in Portugal, you know, the wage difference is not the same as that between Belgium and Romania. So, yes, Portugal suffers from an economic point of view, but I think their wages are not as bad as in Romania. So … not always very motivated! […] There is also the fact that Romanians are more willing to make efforts […] somehow they are used to very harsh working conditions in their country. So, I think that they are inevitably more prepared, you know, to accept everything.* (Private provider)

As emphasized by the above quotations, discrimination can either be based on historical legacies and stereotypes, which disproportionately affect black African women, or be grounded on other alleged characteristics attributed to specific nationalities.

7 The Multiplication of Formal care providers

In Belgium, the multiplication of care providers in the provision of formal home care is the result of both the fragmentation of services, which is a constitutive element of the Belgian old age care regime and concerns first and foremost public providers, and the proliferation in recent years of new—mainly private or semi-private—initiatives.

The fragmentation of public services is a direct consequence of the state regulation, which establishes the profile of workers, their qualifications and the activities allowed within each service. This fragmentation was intensified in the 1980s, when new activities were introduced and care jobs were further differentiated (Gilain & Nyssens, 2001). The clear distinction between each profession is functional to guarantee the professionalization and the quality of services and of jobs, as well as to draw a clear demarcation line between professional work and family care. However, it can be a serious limitation in terms of flexibility and adaptability of services, as confirmed by both private and public providers.

According to the respondents, it is this lack of flexibility that, together with the increased demand for old age care, paved the way for the numerical increase of both private and informal services, the further multiplication of care providers and the diversification of old care provision. Since both private actors and informal caregivers[7] typically offer alternative and complementary services, such as night shifts or live-in services, they are able to fill the gaps and to meet the specific needs of the beneficiaries. The result is that families increasingly combine different services and that public, private, formal and informal caregivers often work together to assure the care of the older person at home.

This has two crucial consequences. The first is that it generates a serious problem of coordination, for both care providers and families. As highlighted by public care providers, contrary to the past, today families have to rely on different services simultaneously and providers have to adapt to this new reality. While in the past a single care provider could meet the demand of the older person, up to his/her transfer to residential care, today to cover the increased demand for home care different types of services are needed, whose cooperation is not always possible, nor easy.

> *It is true that we have serious problems in establishing a collaboration, because at the end of the day those services [private care providers] don't owe us anything, so they are not accountable to us. So, we can ask to make some efforts, we call… when there is something wrong we call the service provider. … We really try to coordinate at the best, but sometimes it works for a week or two and then … well, unfortunately, we don't have any control.* (Public provider)

The fact that each provider is independent from each other makes the collaboration even more challenging, as the exchanges between the different parties cannot be regulated institutionally. The reliance on multiple care providers also requires a considerable effort on the side of families, who spend a lot of time and resources organizing the services. In this sense, despite the strengthening of home care services and regardless of the type of service used by households, the Belgian old age care system increasingly requires the intervention of families in the care of older people, at least in a role of 'care manager' (Ambrosini, 2013).

The second important consequence is that, despite the many advantages linked to the multiplication of actors (such as the availability of a wider range of options), the difficulties linked to their coordination encourage families to increasingly rely on paid informal caregivers. Although none of the respondents felt comfortable with dwelling on the phenomenon of the undeclared work in the sector, all of them acknowledged its increase in recent years, as a consequence of the recent changes occurred at policy level, as well as on the growing demand for care. Since paid informal workers are able to fill the gaps left by formal services and meet the needs of families outside the boundaries of state regulations, they may represent an easier solution for families. The fact that paid informal workers are often recruited through trusted networks contributes to the success of this option.

It is interesting to mention that, instead of being a homogeneous phenomenon, undeclared work in the home care sector takes different forms. As it emerges from the narratives of respondents, two main categories of undeclared caregivers can be identified: the first including workers from formal care services or from other sectors of the labor market, who do extra hours—off the books and often at night—to complete their wages, and the second including caregivers whose main occupation falls under the undeclared economy. In the first case, undeclared caregivers intervene in the care of the older person, together with other services providers. A typical example is a family that recruits several workers to cover 24-hour care, each worker through a different system: one or more workers from public or private providers for a few hours per day, a worker paid under the voucher system for the remaining hours of the day, and sometimes paid with extra cash for additional hours, plus other caregivers paid off the books for nights and weekends. This formula implies that formal providers are in contact—and often work in synergy—with undeclared workers.

> *Actually, we work more and more with the informal network, be that caregivers working in the black market or other. We know that the offer is not enough to cover the demand anyway. [...] So, actually we have many situations where we take over from someone who was there, or vice-versa. This often happens for the night, for example. Actually, there are many people who come to look after the beneficiary for the night, while we are there during the day. [...] Often our caregivers take over from people who work in black.* (Public provider)

> *When it comes to the profession of caregiver, it is clear that undeclared work is extremely developed. [...] There are many nurses who offer services of care assistants ... they do the nights all over the place, for 100/120 euros, sometimes 150, in black, in parallel to their work.*

In the second form, the work is performed on an occupational basis and usually involves longer hours or live-in arrangements. In this case, workers are often the only professionals intervening in the care of the older person and cover all the tasks required, allowing families to save considerable efforts in terms of both time and financial resources. Since the problem of the coordination of services and the management of the different providers is a real burden for families—as mentioned earlier—live-in arrangements clearly represent a simpler option, albeit with other disadvantages (i.e. lack of control on the quality of care and on the working conditions of caregivers, intensification of emotional bonds, etc.).

In both cases, this work is mainly performed by workers with a migrant background. Since the ethnicization of the workforce is visible even in the formal sector, it is not surprising that the informal work—and in particular live-in work—mainly attracts foreign-born women, sometimes with irregular administrative status, who represent the most vulnerable among all migrants. The positioning of migrants in the sector thus seems to take place according to a sort of hierarchy of desirability of professions, at the bottom of which are undocumented migrant women working without a regular contract and residing with older people.

8 Conclusions

This chapter has explored how the recent trends that occurred in the Belgian older care system have influenced the options available to families, on the one hand, and the recruitment of migrant caregivers in both the formal and the informal home care sector, on the other hand. Based on the analysis of in-depth interviews with key informants working in the field of old age care in the Region of Brussels, I argue that the strengthening of home-based services, the growing privatization of care services and the multiplication of actors in the sector, among other factors, have important consequences in terms of families' choices and in terms of the 'ethnicization' of the workforce.

Based on the narrative of the professionals working in the old care sector in Brussels that I have interviewed in the framework of this study, I argue that these recent trends have two crucial consequences for care receivers and families. The first is that families are increasingly encouraged to put in place an individualized 'care mix' that combines public, private, formal and informal services. All the respondents acknowledge this as a new phenomenon: because the offer of public services cannot meet the increased demand, different care providers intervene and interact with each other, often without a formal coordination. Although the strengthening of home care services and the multiplication of care providers do offer a wider range of options to households, respondents emphasize that this new situation demands greater efforts on the side of families, in terms of coordination and in terms of financial resources. This seems to suggest that, as observed in other European countries, also in Belgium the recent reforms in the care provision tend to promote a shift of responsibilities from the state to the families.

Another consequence that the privatization and the multiplication of care providers may have on care receivers and their families is a growing polarization between households that can afford to choose the most adapted solution and those that are bounded to the use of (limited and fragmented) public services, without any other option. As mentioned by most of the interviewees, the fact that even informal solutions are too expensive for a part of the Belgian population seems to suggest that low-income families will have access only to public care services, and only for a limited amount of time, despite the increased variety of care provision. I argue that this raises questions about social justice and the social and economic inequalities between those who can and those who cannot finally enjoy their 'freedom of choice', and that it may ultimately exacerbate social divisions between social groups.

Regarding caregivers, the recruitment of women with a migrant background in home care is considered by all respondents as one of the most visible and growing phenomena in the Region of Brussels. This phenomenon is not exclusive to certain care providers, but concerns the formal—public or private—and informal sectors alike. In line with the situation observed in other European countries, also in Belgium, the reasons for the increased number of women with a migrant background in the domestic and care sector are multiple and include the perceived limit of the supply of locals and the fact that migrants are usually more vulnerable in the labor market, thus more likely to accept lower wages and poorer working conditions (Shutes & Chiatti, 2012).

Based on the elements highlighted by the respondents in my study, I argue that the recent trends in the old age care system have reinforced the likeliness of finding women with a migrant background in care work. Overall, the degradation of the working conditions in the sector, also due to the privatization of services, have contributed to the employment of workers with a migrant background, who represent in Brussels the most vulnerable segments of the labor market. Also, the strengthening and the multiplication of home care services encourage families to find their own solutions, which include the direct employment of migrants working without a regular contract, which may represent a more suitable solution for families in terms of price and coordination.

To conclude, it is important to observe that these trends could potentially lead to an increase in social inequalities, not only between households of different socio-economic backgrounds, but also between workers. Not only the presence of caregivers with a migrant background is increasingly documented in the old age care sector in Brussels—be that formal or informal—but their position in the sector seems to be determined by their characteristics, according to a hierarchy of vulnerability. Just as the diversification in the provision of old age care and the multiplication of care providers leads to a polarization of care jobs, from the more to the least attractive, migrant workers tend to be distributed along a similar ladder, at the bottom of which we find undocumented migrants working under live-in arrangements.

Notes

1. The relatively high expenditure on long-term care and the extensive coverage of residential services for older people are generally associated to social-democratic welfare regimes, which are typical of Northern Europe. However, similar to Bismarckian welfare states, the total budget allocated for social protection is mainly financed by the contributions linked to employment and only partly from general taxation, a feature commonly associated to corporatist welfares (Morel, 2007; Dumont, 2017).
2. Today in the Region of Brussels residential services are mainly private, with more than 78% of private for-profit elderly/nursing homes, which receive a government subsidization identical to that of public facilities (InforHomes, 2017). In the home care sector, government funding only covers non-profit organizations, through a system of accreditations granted by the French and Flemish Communitarian Commissions (COCOF, VGC and COCOM).
3. In this chapter, I use the term 'formal care sector' to indicate all the services—private or public—that are officially recognized by the Belgian state and are thus part of the official labor market. It is used in opposition to the 'informal care sector', which refers to undeclared work (work carried out without a regular contract). Based on this definition, both formal and informal sector refer to paid services and exclude the unpaid care work provided by family members.
4. For French-speaking beneficiaries, two different scales apply, depending on whether the accreditation is granted by the COCOF or the COCOM (for the French-speaking community), but they usually range from 0.50 to 8 euros per hour. A higher scale applies for organizations accredited by the Flemish Communitarian Community (VGC).
5. According to the evaluation report 2018 of the voucher system (*titres-services*), less than 2% of workers are of Belgian origin. Regarding care jobs, the institutional available data does not allow to disaggregate at the level of care jobs (SPF Emploi & UNIA, 2017).

6. Without going into details in the categorization of migrants, a migrant is usually defined as a person born in a country other than that of residence, while second and third generations include individuals with a migrant background (i.e. migrant parents/family) but born in the country of residence—regardless of their current nationality. Since data are not systematically collected by care services providers, in the narrative of the respondents the two categories often overlap.
7. By 'informal caregivers' I intend paid undeclared workers, rather than informal family help.

References

Ambrosini, M. (2013). *Immigrazione irregolare e welfare invisibile. Il lavoro di cura attraverso le frontiere*. Il Mulino.

Anderson, A. (2012). Europe's care regimes and the role of migrant care workers within them. *Population Ageing, 5*, 135–146.

Anderson, B. (2000). *Doing the dirty work? The global politics of domestic labour*. Zed Books.

Anderson, B., & Shutes, I. (2014). *Migration and care labour: Theory, policy and politics*. Palgrave.

Bettio, F., Simonazzi, A., & Villa, P. (2006). Change in care regimes and female migration: The 'care drain' in the Mediterranean. *Journal of European Social Policy, 16*(3), 271–285.

Catanzaro, R., & Colombo, A. (2009). *Badanti & Co. Il lavoro domestico straniero in Italia*. Il Mulino.

Degavre, F., & Nyssens, M. (2012). *Care regimes on the move. Comparing home care dependent older people in Belgium, England, Germany and Italy*. Centre Interdisciplinaire de Recherche Travail, Etat et Société, Université Catholique de Louvain, Charleroi, Belgium.

Dumont, D. (2017). La sécurité sociale et la sixième Reforme de l'état: rétroactes et mise en perspective générale. *Revue belge de sécurité sociale*.

European Commission. (2007). *Health and long-term care in the European Union*. Special Eurobarometer 283/67.3.

Gerhard, U., Knijn, T., & Weckwert, A. (2005). *Working mothers in Europe. A comparison of policies and practices*. Edward Elgar.

Gilain, B., & Nyssens, M. (2001). Belgique: l'histoire d'un partenariat associative – Public. In *Les services sociaux entre associations, état et marché, l'aide aux personnes âgées* (pp. 47–65). La Découverte.
Guo, M., & Willner, S. (2017). Swedish politicians' preferences regarding the privatisation of elderly care. *Local Government Studies, 43*(1), 1–21.
Idea Consult. (2018). *Évaluation du système des titres-services pour les emplois et services de proximité en Région de Bruxelles-Capitale.* Report 2018, Bruxelles Économie Emploi.
Infor-Homes Bruxelles. (2017). *Situation de l'offre d'hébergement pour personnes âgées en maisons de repos bruxelloises.* www.inforhomesasbl.be
Leitner, S. (2003). Varieties of familialism: The caring function of the family. Belgium in comparative perspective. *European Societies, 5*(4), 353–375.
Lutz, H. (Ed.). (2008). *Migration and domestic work: A European perspective on a global theme.* Ashgate Publishing.
Morel, N. (2007). From subsidiarity to 'free choice': Child- and elder-care policy reforms in France, Belgium, Germany and the Netherlands. *Social Policy and Administration, 41*(6), 618–637.
Naldini, M., Wall, K., & Le Bihan, B. (2013). The changing mix of care in six European countries. In B. Le Bihan, C. Martin, & T. Knijn (Eds.), *Work and care under pressure: Care arrangements across Europe* (pp. 171–194). Amsterdam University Press.
Palier, B. (2006). The politics of reforms in Bismarckian welfare systems. *Revue française des affaires sociales*, 47–72.
Pavolini, E., & Ranci, C. (2008). Restructuring the welfare state: Reforms in long-term care in Western European countries. *Journal of European Social Policy, 18*(3).
Ramos, E. (2015). *L'entretien compréhensif en sociologie: Usages, pratiques, analyse.* Armand Colin.
Ranci, C., & Pavolini, E. (2015). Not all that glitters is gold: Long-term care reforms in the last two decades in Europe. *Journal of European Social Policy, 25*(3), 270–285.
Rostgaard, T. (2007). Constructing the care consumer: Free choice of home care for the elderly in Denmark. *European Societies, 8*(3), 443–463.
Rummery, K. (2009). A comparative discussion of the gendered implications of cash-for-care schemes: Markets, independence and social citizenship in crisis? *Social Policy and Administration, 43*, 634–648.
Schwiter, K., Berndt, C., & Truong, J. (2018). Neoliberal austerity and the marketisation of elderly care. *Social and Cultural Geography, 19*(3), 379–399.

Shutes, I., & Chiatti, C. (2012). Migrant labour and the marketisation of care for older people: The employment of migrant care workers by families and service providers. *Journal of European Social Policy, 22*(4), 392–405.

Simonazzi, A. (2009). Care regimes and national employment models. *Cambridge Journal of Economics, 33*, 211–232.

SPF Emploi, Travail et Concertation Sociale and UNIA. (2017). *Monitoring socio-économique 2017: marché du travail et origine*. Online. https://www.unia.be/files/Documenten/Publicaties_docs/1215_UNIA_Monitoring_2017_-_FR-Anysurfer.pdf

Statbel (2017). Structure de la population. Online: https://statbel.fgov.be/fr/themes/population/structure-de-la-population#figures.

Willemé, P. (2010). *The long-term care system for the elderly in Belgium*. ENEPRI Research Report N. 70. ISBN 978-94-6138-012-8.

5

Individual Trajectories and Intersecting Regimes: Methodological Reflections on Researching Migrant Care Work

Bernhard Weicht

1 Introduction

In most developed welfare states, tightening budgets for social expenditure and changing family structures meet continuously increasing care needs. For several decades now, one answer to the arising old age care gaps has been the employment of migrants in both institutional settings and private households (Da Roit & Weicht, 2013; Escrivá & Skinner, 2008; Hillmann, 2005). While migrant work in the domestic sector has been discussed extensively, care in institutions has received less attention. However, also in the formal and/or institutional sector, the positionality of migrant care workers is often precarious, marginalized and vulnerable to exploitation. While we can observe the employment of migrant care workers in many countries with diverse care policies, significant differences in both employment and position are apparent (Lutz, 2008; Flaquer & Escobedo, 2009). Over the last years, a rich body of literature has been

B. Weicht (✉)
University of Innsbruck, Innsbruck, Austria
e-mail: bernhard.weicht@uibk.ac.at

V. Horn et al. (eds.), *The Global Old Age Care Industry*,
https://doi.org/10.1007/978-981-16-2237-3_5

developed, seeking to explain reasons, dynamics and structures of migrant care work patterns. While studies employing a micro focus often zoom in on individual experiences of migrant care workers, macro perspectives attempt to trace the underlying mechanisms producing the very formations and patterns.

Established concepts such as care chains, care regimes or intersections of care, migration and employment regimes (for an overview see Lutz, 2018) explain country-level differences in employment patterns and point to the interrelations of various policy fields in shaping conditions, possibilities and circumstances for both migration and integration within national care systems. While these approaches offer insightful explanations of macro factors shaping individual patterns, they inevitably remain based on the idea of stable macro-micro relationships. However, the consequences of intersecting regimes need to be understood as contingent dynamics following various social mechanisms that might or might not be activated (Porpora, 2015). While this observation seems trivial, identifying when which kinds of mechanisms (i.e. policy intersections) are activated proves rather challenging. This endeavor requires a lens through which individual processes and stories can be followed and through which consequential policy intersections can be identified.

Still aiming to offer an integrated analysis of the social, political and cultural interrelations and intersections that shape the situation of migrant care work, this contribution seeks to explore ways to understand migrant work through the individual trajectories that lead people into the care sector in another country. Starting with the individual narrative allows an exploration of those structures, processes and dynamics that touch and affect these migration and professional journeys. At the micro level, individual trajectories intertwine and intersect—while being enabled, shaped and defined by social, political and cultural systems. These material, cultural and social circumstances, likewise, are produced by an intersection of political, cultural and social structures, as well as people's surroundings, workplaces and institutions. In other words, following the individuals on their narrative journeys provides the possibility to identify the intersecting points at which certain regime dynamics are activated and shaped.

By discussing one case study in detail, I try to show how focusing on individual narratives and trajectories allows an embedded and localized understanding of how particular work and migration patterns come into existence, how they play out within institutional settings and which role individual actions and decisions play in this context.

2 Policies, Regimes and Intersections: Explaining Patterns of Migrant Care Work

In researching migrant care work, much of the literature in sociology, social policy, economics and social care has focused on individual case studies and thus investigated isolated aspects and characteristics, such as employment conditions, care practices, discriminations, or migrant workers' own caring responsibilities (see Anderson & Shutes, 2014). Dyer et al. (2008) describe care work as the combination of emotional labor and bodywork which is poorly paid and has low status and which thus can be seen as both reason and consequence of the recruitment of migrant workers. Fine and Mitchell (2007) emphasize low pay and low status as characteristics of care work. They argue that in this sector class, gender and ethnic power relations are interlinked and that in particular poor women from vulnerable ethnic minorities are employed in the care industries. In a separate strand of literature focusing on the micro level, the attention has shifted to skilled workers in the health care sector (Ahmad, 2005; Triandafyllidou et al., 2016; Yeates, 2009). Due to shortages in the nursing workforce in Western countries, a world labor market has emerged in which nurses migrate, often, but not exclusively, from relatively poor to wealthy countries. Isaksen (2012) thus argues that the national health provisions in various countries are becoming global workplaces. However, many well-trained nurses are not working in the profession they are qualified for but are employed in long-term care homes, in residential homes or as domestic carers (McGregor, 2007). At the same time, care assistants and other personnel in these respective institutions similarly often have a migration background and work side-by-side with those nurses (Cangiano et al., 2009; Doyle & Timonen, 2009; Lyon & Glucksmann, 2008). While the literature has so far differentiated between qualified nurses as health care professionals (Ahmad, 2005; Lowell & Findlay, 2001) and people working in long-term care, Yeates (2009) demonstrates that professional nurses actually form a major category of migrant care workers on all levels. Additionally, as Maroukis (2017) points out, the organization of care work in the receiving countries often allows for upward mobility into increasingly professional sectors of the field.

Combining those micro-level analyses, we can conclude that migrant care workers' positionality (and potential vulnerability) is shaped by interrelated dynamics of both the status of care and people's migration background. Importantly, this group faces the positionalities as a result of their blended status as migrant care workers, a position that cannot be reduced to any of the three components (being a migrant, being a carer and being a worker). Rather, national care arrangements have carved out positions for migrant care workers on all levels. These considerations lead to the question which specific policies foster and hinder the constitution and reconstitution of the very dynamics that shape and change positionality.

On the macro level, it has been argued that the position of migrant workers employed in European care systems is defined and shaped not by a singular policy initiative but by the intersection and interrelation of national social policy and migration regimes (Williams, 2012; Van Hooren, 2012; Da Roit & Weicht, 2013). A combination of political, economic and cultural differences is often described as the reason for different national patterns of employing migrant care workers (Da Roit & Weicht, 2013). Specific social policies, for example, have fostered the employment of migrants in the care sector (Escrivá & Skinner, 2008; Gerling, 2003; Hillmann, 2005; Lutz, 2008; Scrinzi, 2008; Williams & Gavanas, 2008). Beside welfare state policies, company policies, cultural factors and the agency of social actors together form what Pfau-Effinger and Geissler (2005) have conceptualized as 'care arrangement', an explanatory framework for the structuring of care work in a society. Van Hooren (2012, p. 133) emphasizes the role of the social policy sector (in particular in relation to the organization of care) and argues that 'differences in the importance of migrant workers in social care can be explained primarily by differences in social care policies and care regimes'. Similarly, Da Roit and Weicht (2013) identify the availability of public care services as one of the main determinants of the existence of substantial numbers of migrant care workers in a country.

In addition to care arrangements, national migration regimes play an important role in determining migrant work patterns. Sassen's concept of bridges of migration functions as a theoretical building block for the analysis of the migration regime dynamics, that is, its structures and developments over time. A migration regime is understood as the systemic conditions of a country, developing and creating migration patterns over time. Sassen (2007) points out that (interrelations of) historical, economic and political links between countries foster the processes of the push and pull factors of migration. She furthermore shows that pre-existing links such as colonialism or economic domination establish migration bridges through which organized and informal recruitment of workers (through governments, organizations, migrant communities) takes place. As a result, particular emigration-immigration dynamics arise, 'some rooted in past imperial conditions, others in the newer development asymmetries that underlie much migration today' (Sassen, 2007, p. 149). Each of these links represents both structures and actions, or, in Sassen's (2007, p. 163) words, 'the intersection of formal systems and actual practices'.

The different strands of policy regime analyses sketched above have identified important structural influences that determine and affect the different variations of how, where and why migrant workers are employed in national care systems. However, these structural dynamics are only actualized in and through concrete circumstances. Joan Acker's (2000, 2006) concept of inequality regimes as 'loosely interrelated practices, processes, actions, and meanings that result in and maintain class, gender, and racial inequalities within particular organizations' (Acker, 2006, p. 443) helps to situate the concept of positionality within the concrete and specific institutional structures and processes. Acknowledging the case-dependent intersections of care, migration and employment regimes requires analytical embedding. Acker (2006) points to the institutional setting as defining context for inequality between groups of people. In institutions, disparities can exist about the goals of processes, the possession and use of resources and the definition of desirable outcomes. Furthermore, power relations strongly influence specific workplace decisions such as how to organize the work and who enjoys opportunities for promotion and qualified work. However, people's positionality must not be understood as being constructed within independently functioning institutional units either. Rather, institutions as inequality regimes (Acker, 2000) are closely related to societal configurations.

If we accept the idea of context-depending intersecting structures, policies and regimes, an exploration of the relevant intersections needs to start with that unknown context. Rather than predefining the context (e.g. by the reference to national policies and categories), I seek to embed the exploration of structures in people's own journeys across countries and professions. Following the people in question through their trajectory offers the possibility to stop at the respective points at which intersecting policies, regulations and regimes affect the trajectory and the ultimate positionality.

3 Methodological Remarks

The methodological and epistemological reflections presented in this contribution stem from broader analyses of three extensive empirical case studies, carried out in three different European welfare states (Austria, Netherlands and the UK) representing different care, migration and employment regimes (Da Roit & Weicht, 2013). For the empirical work, semi-structured, narrative interviews with in total 65 migrant care workers, accessed through institutions, unions or support networks, were conducted.

The interviews lasted approximately one hour and followed an open narrative guide. The questions for the interviews should trigger people's reflections on their own migration journey (how they came to the country where they are working and living at the moment), their professional journey (how they came to work in care) and their current work circumstances. Interviews were recorded and subsequently transcribed. The analysis followed several steps. The first question for the analysis was which circumstances and processes shape people's positionality? An intersectional approach with a focus on people's social position was chosen. Rather than linking the migrant care workers' situation to identity categories such as class, gender, ethnicity and status, an understanding of the actual situation and its underlying factors depends on the intersection of categories and, importantly, intersections of processes of (mis)recognition. Fraser's (2008) conceptualization of positionality was used to identify the themes, which play a role in shaping people's professional roles. Finally, drawing on Acker's (2000) work, the interviews were coded according to the elements shaping positionality and their manifestations in people's work experiences. The professional position cannot be reduced to abstract socio-economic categorizations in society. Rather, hierarchies of professional status are shaped by normative value systems (Evetts, 2003) within and beyond specific employment settings.

While the claims of this chapter are based on the project's overall analysis, in the following section I discuss one singular case (Mary) in detail. By focusing on the specificities of one person's professional and migration trajectories, I attempt to illustrate the several context-dependent moments in which different elements of policy regimes intersect. Additionally, the example allows a more comprehensive understanding of the role of historical, social and political circumstances that shape policy intersections. The reflections on this particular case set the base for the methodological and epistemological arguments developed in this contribution.

4 Case Study: Mary

4.1 Migration Trajectories

Mary works as care assistant in an old age care home run by a large non-profit organization in central Vienna. When Mary is asked to recall the trajectory of how she came to live in Austria, she starts with her birthplace in former Yugoslavia.

Mary: *I am originally from Bosnia. I was born there. And lived there from 1989 to 92, then there was war, and then we fled with my family to*

> *Serbia, and since my uncle has been living in Austria for 30 years, my father came to him to look for work, because he did not want to be in the war, and we stayed with my aunt in Serbia—there with mother and brother. We went to school, and only after school—father was here all the time—and only after school, in 1999 we came to Austria, because, I wanted to start there, I attended school for qualified nurses, and my father did not want me to stay there, and work for 180 Euros. He said, do you want to come to Austria, as a cleaning lady at the beginning you get 1000 Euros. And so I came to Vienna. And six years as a cleaner in a hotel. … And then married here, found a husband, had two children, and for two years I have been here, in the meantime my diploma recognized, and I now work here.*

In this introductory statement, Mary's recollection of her migration trajectory demonstrates several of Sassen's (2007) denominators of modern migration regimes. While regional conflicts and wars force Mary into the position of a refugee, this trajectory can only be fully understood in its complex local context of Yugoslavia, where ethnic and national relations and differentiations intersect with cross-country family relations. When, subsequently, Mary's migration trajectory takes another turn with her move to Austria, other contributing structures and dynamics override the original push and pull factors. Family relations continue to play a role with the father convincing her to join him in Austria; in order to understand his presence in the country, we additionally have to recognize the long-lasting central European guest worker schemes that previously established existing and utilized migration bridges. Sassen's concept of migration bridges further needs to be localized by taking the particularities of the Yugoslavian (or, at this time already Serbian) school system into account (an aspect I will discuss below).

The analysis of Austria's immigration and employment policies thus needs to employ a historical perspective, which links very different rules and laws to the specific local circumstances. Regulations concerning the guest worker schemes starting in the 1960s (Ivanovic, 2013) interrelate with family reunification rules and immigration regulations after the Yugoslavian wars. Reading these political regulations through the perspective of migration trajectories (see Vaittinen, 2014) allows a connection of historically dispersed social, political and cultural events and decisions that build the migration bridges Sassen describes. Historically, regional links within and between different empires have shaped socio-economic and political realities as well as identities (Batt, 2007). When Mary talks about her experiences (and her self-image) of being a foreigner in Austria these historical continuities become apparent:

Mary: *I know I wasn't born here, but I also know that the other 60% weren't born here either, real Austrians are not born here, they are somewhere from Czech Republic, Slovakia or something like that. And, down there in Serbia, this city was also multi-cultural, very much. From Slovakia I had my best school friend, her name was Schmid, her great-grandfather was in the First World War, the other one was, what's her name, Taubner or something, also down there were many from Romania, Slovakia, Hungary, very, very many, this minority. ... And that's why for me, several languages on the street, nobody says anything, was quite normal.*

At first sight, Mary compares her experiences in Vienna to the situation in Serbia. Underlying her depiction of resemblances, however, lies a historically developed conceptualization of 'Central Europe', with its strong links to the Habsburg Empire (Kuus, 2004; Stenning & Hörschelmann, 2008), which allows discursive analogies (Karner, 2020; Samaluk, 2016) that link Mary's place of residence to her place(s) of origin. The emphasis of historical connections between these regions in Europe and their assumed cultural similarities is a common trope in several interviews I conducted in Vienna. The trope is often used to highlight the connections between countries and peoples and to moderate differences (Hipfl & Gronold, 2011). However, this trope is similarly utilized by politicians in order to more easily demarcate boundaries between different immigrant groups, for example, from the Balkans or Turkey (Hipfl & Gronold, 2011). As Austria's imperial past is emphasized, historical continuities are imagined. Importantly, similar to the description of the migration trajectory, Mary presents these (often idealized) continuities in terms of family relations and regimes.

4.2 Family and Gender Regimes

While I have already pointed to the significance of family relations above, the latter must not be reduced to simple networks or connections between people. Rather, taking a relational approach (see Donati & Archer, 2015), family relations have emergent properties that build broader family and gender regimes. In other words, how families and relationships are conceptualized and lived, in practice, affects the characteristics of migration patterns, in the individual case and as a systemic condition.

I: *And what was your idea at that time, about coming to Austria?*

Mary: *I respect my father very much and, he is a graduated engineer, he also works for 20 years as an assistant here in Austria, he could not, down there, in Serbia. In Serbia we were also refugees, foreigners and so on, even now after 20 years, and he couldn't find himself down there. Here he stayed, we had everything back then, where others had very little. We went through everything: school, excursions, everything. And he has everything, he could offer us everything. And then, when he said, "you want to come to Austria? If you don't want to go on to school, better that you come to me!" I said, "ok, father", that was self-evident when he asked something like that. My parents are not divorced, they are in a marriage. Only, mother works in a pharmacy, as a pharmacist, and she did not want to come to Austria to clean, she wanted to stay in her profession, and she has been working in a pharmacy in Serbia for 20 years. That's why she never wanted to come here, with language and everything, and father works here in Austria in a big company, and he is anyway, I can't say, satisfied, we had everything, and a few more years, then he retires anyway, then he goes back to Serbia. We had a house, built it, rebuilt it. I think he is satisfied anyway.*

Mary narrates the various interrelated situations that led her to joining her father in Austria. Family relations prove crucial for Mary's decision but they are intertwined with the situated context of family practices and their embeddedness in gender relations and national gender regimes.

Due to the experiences of war and subsequent migration, Mary's father accepted a process of deskilling, as he has worked in low-skill jobs in Austria despite his training and experience. Mary partly justifies this with the fact that her father could provide well for his family, thereby maintaining their relatively well-off middle-class position. A classical trope in the context of European labor migration (Bolzman & Bridji, 2019) is the topic of return migration. In contrast to Mary's father, her mother, working as a pharmacist, refused a possible deskilling process by remaining in Serbia. This decision also hints at interesting intersections of class and gender relations and the significance of (professional) women's work. In a context in which state policies promoted women's professional roles in the labor market while at the same time caring roles remained highly gendered (Massey et al., 1995), the arrangements of Mary's parents invite us to delve deeper into an analysis of the particularities of the gender regime in former Yugoslavia.

Looking closely at the several structural influences on Mary's decision to seek work in Austria, we see that fixed categories that usually describe migration processes, such as labor migration, being a refugee or family reunification, only unsatisfactorily capture the actual decision-making process. Moreover, looking at the interrelations of these categories on a family level points to the significance of not only the policy regimes in the receiving but also in the sending country. In other words, the constitution of gender, family, labor and migration regimes in the sending country builds the migration bridge with the respective regimes in the receiving country. The migration decisions of Mary's family are fostered by the connections, symmetries and asymmetries that bind the various regimes together. Starting the analysis from a trajectory perspective enables an identification of the points of connections and intersections between sending and receiving regimes and cultures.

4.3 Professional Trajectory

The migration trajectory allows an identification of different elements affecting decision and process of migration. The question of why people work in care in another country is also based on someone's professional trajectory. Mary found herself in the process of becoming a qualified nurse when she came to Austria, even though this fact did not directly contribute to her migration decision. Rather, her qualifications once again were influenced by the local structures in Yugoslavia/Serbia, more specifically by the educational regime and her own family cultures.

I: *And how did it actually come about that you went into nurse training?*

Mary: *That was my wish somehow. I was always like that. I wanted to take care of everyone. That was … but father wanted me more, that I, I was also good in geometrical drawing, mathematics, physics, that was my strength. And he wanted me to be a mechanical engineer, he always wanted me to go to this side. And somehow through mother, she was in the pharmacy all the time, we were always visiting when we were little. And then … I went to the ambulance with her, I don't know, I was fascinated by this, hospital and medicine. And so, that was my wish. It wasn't like mom's, but my wish. Father was not so enthusiastic. And then brother came and became a physiotherapist, without medicine, he didn't want that. That's how it happened.*

I: *And did you know anyone who worked in the field?*

Mary: *No, no. It all came about, I don't know, somehow. At that time—so that my father remains calm—I also … I enrolled in a mechanics school for technical training, in a medical school and another graphic design school. There you could choose more schools, there was an entrance exam, but for all schools at the same time, and then later you could choose which school you would be admitted to, you could choose which school you would attend. And I already enrolled, and there I saw that I was accepted at the medical school. Then I decided to go there. I wanted to go there, I lived in a small town in Serbia,*

> *not small, but 300,000 inhabitants or so, and there was no higher school for nurses, it was only in the capital, in Belgrade, and my father, he didn't want me to go to the medical school, that I go to Belgrade, alone, and that he finances everything, I wanted that, he didn't want that, and then I tried, a year, I studied technology, but I was not satisfied, and dropped out, and then I wanted to start working in [city], I also wrote applications, everything, and he came home on weekends … I say I want to work—no, you're not going to work, you come with me [laughs]. And so, really, we did not discuss it. I said fine, I'll come with you. [laughs]. No problem.*

Mary's recollection once again centers on family relations. In retelling her professional trajectory, we see the embeddedness of family considerations within local structures and contexts. Both her mother and her father promote professional roles and thus a particular class situation by encouraging Mary to follow her own interests and choices. This ideal necessitates the availability and accessibility of various possible professional trajectories so that choices can be executed. Her family's class background therefore intersects with the educational system in which (nursing) care was presented as one option, equal to other professional trajectories. Zaviršek (2008) shows that the professionalization of social work in Yugoslavia started originally with a male rather than female majority and that this gender order was not only a consequence of a strategic structural focus but also a counter move away from the pre-war tradition of female charity and philanthropy. Social work and (nursing) care work have thus been framed as professional career options, often much earlier than in possible receiving countries. Through specific migration regulations, Austria has been able to utilize the geopolitical turmoil in the region to fill labor shortages with highly skilled refugees. Many care workers in my sample, in particular from Slovakia and the former Yugoslavian countries, refer to the comparatively high standard and status of their education. The socialist professionalization of care was technical and medical, which leads to important gender dimensions. The (post) socialist training to becoming a nurse follows a medical logic and is often functioning as prerequisite for several medical professional trajectories.

Care workers themselves clearly recognize the high level of skills and knowledge they possess, which might distinguish them from other colleagues in their work places in the receiving countries.

In the second part of the extract above, Mary recalls her choice for a medical trajectory and its interrelations with possible migration perspectives. Continuing in this professional direction required her to migrate: in the case of the medical school to the capital, or, together with Mary's father, to Austria. In both cases, the migration trajectory can be seen as professional advancement, even though the actual professional positions and opportunities differ. Contrary to the earlier mentioned binary between professional, skilled health care migration and finding work in long-term care as an ultimate possibility for migrant workers, Mary's case illuminates the possibility of utilizing existing migration bridges as professional development. Mary dedicated her trajectory to care in Serbia, though this choice was not a reason for the migration process itself. Rather, the interrelation of regime structures and established migration bridges normalized migration as an option of personal (and professional) choice.

4.4 Deskilling Regimes of Migrant Care Work?

A final trope often explored in the global migrant care literature (e.g. see Cuban, 2013) is the process of deskilling, in particular when qualified (health) care workers are employed below their qualification level. Looking at Mary's case we can note the importance of regime intersections in that employment and migrant labor regulations intersect with professional structures and arrangements in the field of care, both in Austria and in her country of qualification (Serbia). While zooming in on her particular experiences, however, another feature appears to challenge the clear-cut scrutiny of different employment levels: the cultural constitution of care arrangements (Pfau-Effinger & Geissler, 2005).

Mary: *I've only recently come into contact with geriatric care, because down there is not so usual. Down there are very, very few nursing homes, and in my internship I was in hospitals, and I was really a real nurse, technical-medical, and I didn't have so much [?], with this relationship building and things like that. I was really, really outgoing and all that stuff. And only here, I can see that I'm also a nursing home type, I, I didn't think I was like that before.*

I: *And why?*

Mary: *Atmosphere does me good. Like at home. Somehow. Since I do not have so strict, disinfected environment, so, it feels like home. When I come here, I don't feel like I'm at work, I really feel like home. And in the hospital, I also had an internship here, everything is strict, disinfected ... also with residents, some come and stay only 2 or 3 days, go home. You can't make that connection. ... And it's not, it's not for me.*

Mary's evaluation shows different 'cultures of emotional labour' (Dyer et al., 2008) which she retrospectively links to differences in her practices of nursing between her country of origin and the country of residence. This does not mean that national cultural practices are the main causal explanation; rather, the migration bridge is utilized to make sense of and justify different professional positions through cultural explanations.

The need to perform emotionally demanding care work rather than technical and medical tasks can be particularly challenging for migrant workers who originate from places with different professional identities in care. The devaluation of skills, knowledge and the professional role might be particularly difficult for migrant workers coming from places where their skills and qualifications have a high symbolic value (Ramirez & Rainbird, 2010; Keep & Mayhew, 2010). In contrast, the (necessarily incomplete) contrasting of high-status health care on the one hand and low-status long-term care on the other can also lead to a positive revaluation of one's positionality, as exemplified in Mary's case. While avoiding falling into the trap of romanticizing deskilling processes after migration (see Bolton & Muzio, 2008), it can be argued that different national care arrangements potentially offer different professional trajectories.

5 Conclusions: Intersecting Regimes—Intersecting Trajectories

The interrelation of care, employment and migration regimes marks one of the main strands in the conceptual and empirical literature on explaining migrant care work patterns. The interplay of different policy fields and socio-political arrangements is argued to define and foster the particular configurations of where and how migrants are employed in national care systems. However, macro-level explanations not only often favor structure over process but they inevitably assume consistent social mechanisms and processes at work. Whether and how intersecting policies shape migrant care workers' positionalities are context-dependent questions which is why I particularly wanted to focus on those contextual elements.

The effects of policy intersections are only activated under circumstances and can, likewise, only be utilized under certain contextual conditions. However, if context-dependency is crucial for an understanding of regime structures, how can we identify the relevant context in which an analysis needs to be embedded? Above I propose to look at people's individual trajectories, told from their own recollections. Relying on narrative, biographical interviews by definition always means a reliance on people's own chronicles of their lives. This includes accepting that a participant's narration is a personal selection of telling their life choices, developments and trajectories. Retrospectively, some elements appear more important than others, some aspects or dynamics have been forgotten, while others are told in a more coherent way than they were originally experienced (Riessman, 2008). However, if we want to understand the logics of how professional and migration trajectories intersect in building care migration bridges, it seems inevitable to start with people's own logics of their experiences. In their narratives, participants establish more or less coherent trajectories that provide insights into the actual significance of regime characteristics. Carefully exploring personal trajectories allows a demarcation of those points at which different regimes intersect and how those intersections shape migrant care work. Drawing on the example above, I suggest four steps in this approach:

Firstly, starting with the individual stories, Mary's narrative shows that professional and migration trajectories intersect at many points. Mary's professional route toward care, for example, was altered through her migration practices and, ultimately, redefined in the light of her experiences with long-term care. Secondly, the ways trajectories intersect need to be understood as embedded in the respective political, social and cultural regimes (both in the sending and in the receiving countries). By starting with exploring people's migration and professional trajectories, I was not bound to an analysis of typical regime structures. The case above demonstrates, for example, how national gender regimes can influence work and migration decisions, which in turn link to other regime aspects and contexts. The role of national educational systems or the specificities of the interethnic family and migration relations within former Yugoslavia are further illustrations of this point. Thirdly, the resulting trajectories are embedded in the complex interrelations of receiving regimes, where, for example, the national conceptualization of long-term care or historical guest worker schemes appear significant. Finally, regime intersections in the countries of origin interrelate with regime intersections in the receiving countries and thereby actually build and rebuild bridges of migrations.

Starting with people's own trajectories provides a way into the roots and routes of migrant care work. Mary's narrative has demonstrated that different elements of regimes and policies are contingent and are only activated under certain circumstances (as, e.g., the family patterns emerging from guest worker schemes in the 1970s). The case study shows clearly developed migration bridges, based on local labor migration schemes and built upon family connections. Likewise, the configurations of migrant care work, or, on the individual level, people's positionality, need to be understood as based in specific contextual conditions (as the question of defining care and deskilling processes has shown).

For (the intersection of) regime analysis, these steps provide a more grounded, empirically and historically situated examination. We have seen in Mary's account that each change in the personal migration and professional trajectory can affect the course of many other decisions. The ways these vital conjunctures (Liversage, 2009) play out is thus dependent on the bridges and links available.

References

Acker, J. (2000). Revisiting class: Thinking from gender, race and organizations. *Social Politics, 7*(2), 192–214.

Acker, J. (2006). Inequality regimes: Gender, class, and race in organizations. *Gender & Society, 20*(4), 441–464.

Ahmad, O. (2005). Managing medical migration from poor countries. *BMJ, 331*, 43–45.

Anderson, B., & Shutes, I. (Eds.). (2014). *Migration and care labour: Theory, policy and politics*. Palgrave Macmillan.

Batt, J. (2007). Introduction: Defining Central and Eastern Europe. In S. White, J. Batt, & P. Lewis (Eds.), *Development in Central and East European politics* (pp. 1–19). Palgrave Macmillan.

Bolton, S., & Muzio, D. (2008). The paradoxical processes of feminization in the professions: The case of established, aspiring and semi-professions. *Work, Employment & Society, 22*(2), 281–299.

Bolzman, C. A., & Bridji, S. (2019). Older immigrants living in Switzerland and ambivalence related to return around the retirement period. *International Journal of Comparative Sociology, 60*(1–2), 14–36.

Cangiano, A., Shutes, I., Spencer, S., & Leeson, G. (2009). *Migrant care workers in ageing societies: Research findings in the United Kingdom*. Executive Summary. COMPAS.

Cuban, S. (2013). *Deskilling migrant women in the global care industry*. Palgrave Macmillan.

Da Roit, B., & Weicht, B. (2013). Migrant care work and care, migration and employment regimes: A fuzzy-set analysis. *Journal of European Social Policy, 23*(5), 469–486.

Donati, P., & Archer, M. (2015). *The relational subject*. Cambridge University Press.

Doyle, M., & Timonen, V. (2009). The different faces of care work: Understanding the experiences of the multi-cultural care workforce. *Ageing & Society, 29*, 337–350.

Dyer, S., McDowell, L., & Batnitzky, A. (2008). Emotional labour/body work: The caring labours of migrants in the UK's National Health Service. *Geoforum, 39*(6), 2030–2038.

Escrivá, A., & Skinner, E. (2008). Domestic work and transnational care chains in Spain. In H. Lutz (Ed.), *Migration and domestic work: A European perspective on a global theme* (pp. 113–126). Ashgate Publishing.

Evetts, J. (2003). The construction of professionalism in new and existing occupational contexts: Promoting and facilitating occupational change. *International Journal of Sociology and Social Policy, 23*(4/5), 22–35.

Fine, M. D., & Mitchell, A. (2007). Review article: Immigration and the aged care workforce in Australia: Meeting the deficit. *Australasian Journal of Ageing, 26*(4), 157–161.

Flaquer, L., & Escobedo, A. (2009). The metamorphosis of informal work in Spain: Family solidarity, immigration and development of social rights. In B. Pfau-Effinger, L. Flaquer, & P. Jensen (Eds.), *Formal and informal work in Europe. The hidden work regime*. Routledge.

Fraser, N. (2008). *Scales of justice: Reimagining political space in a globalizing world*. Polity Press.

Gerling, V. (2003). Die britische und deutsche Antwort der Altenhilfe auf zugewanderte Senioren/innen im Vergleich. *Zeitschrift für Gerontologie und Geriatrie, 36*(3), 216–222.

Hillmann, F. (2005). Migrants' care work in private households, or: The strength of bilocal and transnational ties as a last(ing) resource in global migration. In B. Pfau-Effinger & B. Geissler (Eds.), *Care arrangements in Europe—Variations and change*. Policy Press.

Hipfl, B., & Gronold, G. (2011). Asylum seekers as Austria's other: The re-emergence of Austria's colonial past in a state-of-exception. *Social Identities, 17*(1), 27–40.

Isaksen, L. W. (2012). Transnational spaces of care: Migrant nurses in Norway. *Social Politics, 19*(1), 58–77.

Ivanovic, V. (2013). The employment of Yugoslav workers in Austria in the 1960s and 1970s. *Zeitgeschichte, 40*(1), 36–48.

Karner, C. (2020). 'Der Balkan' in Der Krone: Austria between 'frontier orientalism' and amnesiac nationalism. *Interventions, 18*, 1–19.

Keep, E., & Mayhew, K. (2010). Moving beyond skills as a social and economic panacea. *Work, Employment & Society, 24*(3), 565–577.

Kuus, M. (2004). Europe's Eastern expansion and the reinscription of otherness in East-Central Europe. *Progress in Human Geography, 28*(4), 472–489.

Liversage, A. (2009). Vital conjunctures, shifting horizons: High-skilled F immigrants looking for work. *Work, Employment & Society, 23*(1), 120–141.

Lowell, B. L., & Findlay, A. (2001). *Migration of highly skilled persons from developing countries: Impact and policy responses*. International Migration Papers. Geneva: International Labour Office. http://nbn-resolving.de/urn:nbn:de:0114-fqs0603143

Lutz, H. (Ed.). (2008). *Migration and domestic work: A European perspective on a global theme*. Ashgate Publishing.
Lutz, H. (2018). Care migration: The connectivity between care chains, care circulation and transnational social inequality. *Current Sociology, 66*(4), 577–589.
Lyon, D., & Glucksmann, M. (2008). Comparative configurations of care work across Europe. *Sociology, 42*(1), 101–118.
Maroukis, T. (2017). Migrant care workers' trajectories in a familistic welfare regime: Labour market incorporation and the Greek economic crisis reality-check. *Journal of Ethnic and Migration Studies, 43*(1), 1–17.
Massey, G., Hahn, K., & Sekulić, D. (1995). Women, men, and the 'second shift' in socialist Yugoslavia. *Gender & Society, 9*(3), 359–379.
McGregor, J. (2007). Joining the BBC (British Bottom Cleaners). Zimbabwean migrants and the UK care industry. *Journal of Ethnic and Migration Studies, 33*(5), 801–824.
Pfau-Effinger, B., & Geissler, B. (Eds.). (2005). *Care arrangements in Europe—Variations and change*. Policy Press.
Porpora, D. (2015). *Reconstructing sociology: The critical realist approach*. Cambridge University Press.
Ramirez, P., & Rainbird, H. (2010). Making the connections: Bringing skill formation into global value chain analysis. *Work, Employment and Society, 24*(4), 699–710.
Riessman, C. (2008). *Narrative methods for the human sciences*. Sage.
Samaluk, B. (2016). Migration, consumption and work: A postcolonial perspective on post-socialist migration to the UK. *Ephemera: Theory & Politics in Organization, 16*(3), 95–118.
Sassen, S. (2007). *A sociology of globalization*. Norton.
Scrinzi, F. (2008). Migrations and the restructuring of the welfare state in Italy: Change and continuity in the domestic work sector. In H. Lutz (Ed.), *Migration and domestic work: A European perspective on a global theme* (pp. 29–42). Ashgate Publishing.
Stenning, A., & Hörschelmann, K. (2008). History, geography and difference in the post-socialist world: Or, do we still need post-socialism? *Antipode, 40*(2), 312–335.
Triandafyllidou, A., Isaakyan, I., & Schiavone, G. (Eds.). (2016). *High-skill migration and recession: Gendered perspectives*. Palgrave Macmillan.

Vaittinen, T. (2014). Reading global care chains as migrant trajectories: A theoretical framework for the understanding of structural change. *Women's Studies International Forum, 47*(B), 191–202.

Van Hooren, F. (2012). Varieties of migrant care work: Comparing patterns of migrant labour in social care. *Journal of European Social Policy, 22*(2), 133–147.

Williams, F. (2012). Converging variations in migrant care work in Europe. *Journal of European Social Policy, 22*(4), 363–373.

Williams, F., & Gavanas, A. (2008). The intersection of childcare regimes and migration regimes: A three country study. In H. Lutz (Ed.), *Migration and domestic work: A European perspective on a global theme* (pp. 13–28). Ashgate Publishing.

Yeates, N. (2009). *Globalizing care economies and migrant workers: Explorations in global care chains*. Palgrave Macmillan.

Zaviršek, D. (2008). Engendering social work education under state socialism in Yugoslavia. *British Journal of Social Work, 38*(4), 734–750.

Part II

Promoters and Intermediaries

6

Self-regulation in a Grey Market? Insights from the Emerging Polish–German Business Field of Live-in Care Brokerage

Simone Leiber and Verena Rossow

1 Introduction

Demographic change and societal aging are challenging EU member states on solutions for the 'care question'. This particularly concerns countries where the family has traditionally been central for care provision and where most dependent older people are still being cared for in their own homes and by their relatives.[1] Who provides care for older people when women are increasingly participating in the labor market and working environments have become more demanding by requiring employees to be flexible

S. Leiber (✉)
University of Duisburg-Essen, Essen, Germany
e-mail: simone.leiber@uni-due.de

V. Rossow
Institute for Economics, Labour and Culture (IWAK), Goethe University Frankfurt, Frankfurt, Germany
e-mail: rossow@em.uni-frankfurt.de

V. Horn et al. (eds.), *The Global Old Age Care Industry*,
https://doi.org/10.1007/978-981-16-2237-3_6

regarding their availability and working times? Due to gaps in national public care systems (for Germany: cf. Theobald & Hampel, 2013), private households increasingly use live-in care arrangements based on intra-EU care migration. Since the EU's eastern enlargement in 2004, the legal framework for this particular kind of labor mobility between East and West is largely determined by the European Single Market.

The phenomenon of live-in care work[2] by migrants, resulting also in transnational 'care chains' (Hochschild, 2000), is not new and well researched from different sociological angles at the macro, meso, and micro levels (for an overview of the main literature see, e.g., Krawietz, 2014, pp. 14–21; Lutz & Palenga-Möllenbeck, 2010, 2015). What is new, however, is the mushrooming of private brokering agencies in this field since the EU's eastern enlargement. Especially in Germany, these agencies have spread considerably and have become, as we argued earlier, new players in a Europeanized care market (cf. Leiber & Rossow, 2016). As we will show, the field is characterized by a high degree of legal insecurity due to legal complexities and enforcement gaps in the EU multi-level system. In addition, the domestic sphere, where this kind of 'intimate labor' (Boris & Parreñas, 2010) is performed, is hard to control and is prone to abuse, hence the reason why we speak of a grey market.

In this chapter, we seek to analyze recent developments in this evolving business sector linked to a grey care market. Our study aims to contribute to research on mapping such private labor market intermediaries in their role of shaping domestic (care) work arrangements. At the same time, we explore how the 'corporatization of care' (Farris & Marchetti, 2017)—that is, where for-profit actors enter the care sector and begin to apply their rationalities and corporate logics—plays out in a grey market. Our aim is to analyze and critically reflect the self-proclaimed 'quality' approaches of such live-in care brokerage companies in relation to the question of decent care work. We do this along the following three dimensions that emerged from our empirical data (see Sect. 3 below). In how far do agencies as intermediaries make a difference: (1) in terms of rule abidance and rule transparency in a highly complex legal environment (coping with the legal framework dimension); (2) in terms of enhancing the professional and interpersonal skills of migrant carers (qualification dimension); (3) in terms of business procedures to reconcile expectations and interests between German customers and Polish workers (intermediation dimension)?

Quantifying live-in migrant care workers (LIMCs) as well as brokerage corporate actors is difficult as it usually concerns shadow or semi-formal economies. In Germany, due to the informal character of the work and the lack of any central registration, there are therefore only estimates that start from an assumption that eight percent of the officially registered care households in Germany employ (in its broadest meaning) an LIMC, meaning that estimates begin at around 160,000 households (Hielscher et al., 2017, p. 95), and reach up to approximately 450,000 LIMCs (Benazha et al., 2021, p. 26).[3]

There is also no reliable information available on the number of brokerage agencies. Analyses based on the screening of agency websites already in 2008 found approximately 70 agencies in Germany. What needs to be taken into account, however, is that some agencies seem to use several different websites to attract customers (for details cf. Krawietz, 2014, pp. 41–47). Nevertheless, the number of agencies has meanwhile certainly increased, and brokerage fees, when made transparent, are on average around EUR 1000 per year (Emunds, 2016, p. 43). In the EuroAgencyCare project,[4] an extensively structured web research strategy was pursued, identifying the internet presence of 274 independent agencies in 2017 in Germany (Leiber et al., 2019). Similar numbers have been presented by Stiftung Warentest (2017; 266 German agencies), Benazha et al. (2021, p. 26) even speak of around 400 agencies. Despite this, black market live-in care arrangements have not disappeared, but seem to prevail in a large part of German households.

State of the Art and Research Gap

With regard to transnational care chains, brokering agencies all over the world occupy a central position within recruitment schemes: 'The employment agencies, which recruit and place domestic workers across national boundaries, are crucial actors in the construction, maintenance and reproduction of global care chains, in which women from the South migrate to the North in order to provide domestic work' (Fudge & Parrott, 2014, p. 70f). Brokerage companies are part of a global trend depicting a growing involvement of temporary staffing firms in labor market topographies (Coe et al., 2010; Eichhorst et al., 2013) and are more generally termed labor market intermediaries (LMIs) (Benner, 2003; Peck & Theodore, 2002; Shire et al., 2018). In short, they seek profitability out of the institutionalization and standardization of labor migration. In the EU context, research has only recently begun to address the phenomena, due to the increase in live-in care migration after the EU's eastern enlargement.

Labor market intermediaries in the live-in care market in Germany are part of a transnational business network which spans at least two countries: one sending (and recruiting) country and one brokerage country (Leiber et al., 2019, p. 374). Apart from very few exceptions, these private companies are profit oriented. Private intermediaries have become crucial players by strongly influencing the conditions under which LIMCs are brokered, employed, and/or supported during their stay abroad. By offering allegedly 'legal' models and, to a greater or lesser extent, comprehensive legal support and contracts, '[t]hey serve as relatively formal bases within a highly informal, invisible market' (Lobel, 2001, p. 96). In Germany, over recent years, a growing market share of entrepreneurial actors have joined together for the common purpose of creating a 'quality segment' in the market.[5] They did so by establishing business associations and investing in standardization procedures and legal expertise, as well as by political lobbying, with the overall goal of draining the illegal market.

To date, most research in the field of German live-in care settings has been aimed at explaining the phenomenon of live-in care work, mostly addressing LMIs only indirectly. Initial research on the phenomenon of live-in care work took a migration or gender perspective (Lutz, 2008), often based on biographical approaches (Karakayalı, 2010), with working conditions being central. Only recently has published research taken professional brokers into account (Krawietz, 2014; Schmidt et al., 2015; Aulenbacher et al. 2021). Krawietz (2014) gave a first and comprehensive overview of the intrinsic logic of live-in care brokerage companies, both in Germany and in Poland, stating that the majority of the German-based companies preferred operating in 'hidden spaces' (2014, p. 91) and avoiding public attention. This last aspect seems to have now changed, at least to a degree. As a general assessment of the relevance and role of brokerage firms, Krawietz concludes by questioning the positive (strengthening) effect of contracts introduced by labor market intermediaries for workers. This effect ends at the point where workers are left to themselves in a private household (2014, p. 151ff; for the Netherlands: Van Bochove & Zur Kleinsmiede, 2019).

However, little has been written on how LMIs in this field might interfere in or influence working and care conditions. Brokerage agencies are under general suspicion of deliberately making money out of exploitative labor relations (Emunds, 2016; Schwiter et al., 2014). Equally, although emphasizing 'legality' as an allegedly 'quality' statement in the narratives of brokerage firms in Germany (Steiner et al., 2019) is emblematic of this market, this field is far from being subject to thorough regulation, and rule compliance being minimally enforced (Böning & Steffen, 2014). And yet this reading seems to prematurely exclude the possibility of substantial momentum on this market, a momentum that finds expression not only in the strategic political positioning (Leiber et al., 2019) of evolving business associations and their members, but also in a publicity offensive strategy (BHSB, 2019; VHBP, 2015) and an interest in stripping off the dodgy image of a shadow sector (BHSB, 2015; VHBP, 2017).

To date, research is absent on the market composition in this field offering a mapping of the presumed variety of entrepreneurial approaches, but it would be too simple to dismiss potential movement and innovation in the field. Yet, profound empirical research is difficult in this grey market, and the fact that this kind of care work is being performed in private households brings with it additional difficulties. There are several qualitative studies selectively touching on the agencies' role from the LIMCs' perspective (e.g. Ignatzi, 2014). Some LIMCs seem to see the agencies as a source of protection and security, others rather as a modern form of exploitation. Nevertheless, difficult data access means that '[w]hat remains in question is the impact these intermediaries may have on working conditions in the sector' (Fudge & Hobden, 2018, p. 2). The same is true for their relevance in terms of 'decent care (work)' (e.g. ILO, 2018).

Method

In the following, we present results from the EuroAgencyCare project that used a mixed methods approach based on a quantitative online survey for agencies combined with 28 qualitative expert interviews with agency representatives in Germany (as well as Poland and Austria) and other key actors in the field. The interviews were based on semi-structured interview guidelines, and the interviewees were heads of agencies and their associations in Germany and Poland, political stakeholders at EU and national level, as well as representatives of unions, civil society, and welfare organizations.

After a first analysis of the online survey results, a sampling strategy for complementary expert interviews with heads of companies was developed. The goal was to cover a range of different companies along the criteria size (in terms of the number of placements per year and number of employees), the kind of service provider (private, for-profit company vs. welfare association), foundation year, scope of service provision (regional vs. national), and political engagement (member of one of the (back then) two business associations VHBP or BHSB: yes or no). According to these benchmarks, seven German agencies were chosen and contacted in order to schedule an interview date.[6] The interviews lasted from one to two hours, taking place in company offices or conference rooms. One interview took place in the private house of a company head. All interviews were audio recorded, verbatim transcribed, and coded using the software MaxQDA. The framework analysis method (Dunger & Schnell, 2018) was used to analyze the interview material. This analytical approach is characterized by a strong degree of formalization and intersubjective replicability by the research data being neatly arranged into matrix tables and gradually condensed into increasingly incisive research theses.

In this chapter, we present project results based on our qualitative data. It is important to note that the data presented here are mainly taken from the head of companies' self-descriptions of their business models and practices. We have to take into account that these interviewees are likely to be interested in a positive presentation of their company. For this reason, critical statements about the agencies' activities are probably rather rare while positive aspects might be overemphasized. Thus, this study should be seen as first approach to a so far almost unexplored field of research where due to the characteristics of a 'grey market' data are generally difficult to access. It aims to display and critically discuss the agencies' perceptions on quality aspects in the field of live-in care; it does not claim to scrutinize in depth what is 'really' going on at the household level.

Based on this material, we aim to contribute to the research discussion on the role of private agencies as intermediaries in live-in care brokerage. In Sect. 2, we show how features of the multilevel regulatory system, legal uncertainty, and lack of European and national enforcement (capacity) have created new opportunity structures for private market actors. In Sect. 3, we map and critically reflect on the self-proclaimed 'quality' approaches of live-in care brokerage companies based on our qualitative interview study. In the concluding Sect. 4, we argue that to enhance decent care (work) conditions, state regulation, particularly at the national level, is needed because the self-regulation of private companies is insufficient.

2 Live-in Care Migration in the EU: Regulatory Features of the Multilevel System

When studying the phenomenon of transnational care migration, we need to take into account the intersection of different legal (migratory, labor market, and care) regimes (cf. Williams, 2010; Wagner, 2015) in at least two countries (here: Germany and Poland), and at different governance levels (international, EU, national, corporate, and domestic). In addition, domestic work takes place in the private domain of the family, which is particularly protected from state intervention and control. Without adequate control and enforcement, however, working conditions in private households are prone to laws being abused.

Considering that LIMC form a group of workers who are not clearly addressed as legal subjects or a specific labor force, their status, their work, and the conditions of their brokerage are only indirectly affected by existing legislation. In this, as Fudge (2011, p. 242) puts it: 'Paid domestic work transgresses a number of boundaries that law both defines and reflects, and it is this transgression, historically, that has made protecting this important activity difficult.' Looking at existing regulatory efforts, Fig. 6.1 illustrates the various regulatory benchmarks within a multilevel

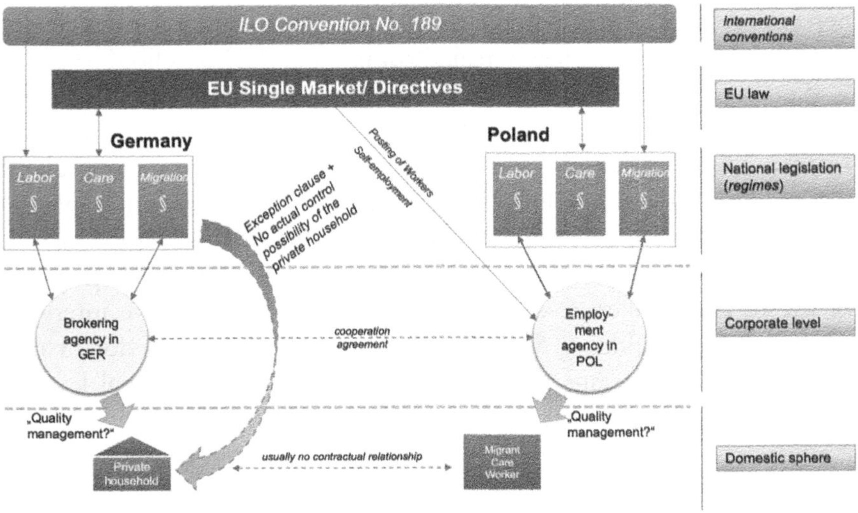

Fig. 6.1 Brokering and employment agencies in the multilevel governance system

political system. The following section outlines the different levels of and (limited) regulatory influence on live-in migrant care work in Germany.

At the international level, on 20 September 2013, Germany ratified ILO Convention No. 189 on the effective protection for domestic workers but used the exception rule in Art. 2 II C189 to exclude the group of persons subject to § 18 I Nr. 3 ArbZG[7] from the scope of this convention (Scheiwe & Schwach, 2013). This step was justified with the claim that it would be impossible for some employees to differentiate between work and recreation times. The applicability of this exemption clause to LIMC is contested (for details see Benazha, 2021). The German Federal Government is aware of this situation but has been acting very cautiously in this respect and simply refers to the responsibility of EU law (e.g. Deutscher Bundestag, 2012).

In addition, the legal frame for recruiting Central and Easter European (CEE) migrant domestic care workers in Germany is largely determined by European regulation. On an EU level, the issue of posting workers has been on the agenda for at least two decades. In the mid-1990s, Directive 96/71/EC of the European Parliament and of the Council of 16 December 1996 concerning the posting of workers in the framework of the provision of services (henceforth Posted Workers Directive: PWD) was introduced with the aim of fostering minimal labor standards for posted workers (e.g. maximum work periods, minimum paid annual holiday, health and safety at work, and more). Likewise, this directive was originally developed to ease professional stays abroad for employees in existing working arrangements. However, it remained fairly general in its regulatory ambitions on labor standards, merely pointing at a general level of political desirability and omitting questions of enforcement. In addition, there is no specific indication on how to deal with the particularities of the domestic sphere. This rather vague regulatory framework, combined with the EU principles of free movement of services, constitutes the background for a 'posting-based brokering model', which German brokering agencies often rely on.

The main appeal to German households to use brokerage agencies (as opposed to using informal black market structures, or becoming formal employers themselves) lies in the low levels of time and effort for customers, and a remarkably fast and flexible provision of domestic workers under a presumed legal frame. In this model, German agencies rely on foreign partner companies to establish a sustainable system for a steady supply of labor. Foreign employment agencies recruit labor in their country of origin and, for a defined period, post them abroad. Social insurance contributions are paid in the country of origin (for the first 24 months), just as wages. Social insurance contributions are generally much lower in the countries of origin than in the destination countries. It is this wage difference that makes the posting of workers interesting and profitable for the sending employers (Wagner, 2015, p. 339).

As regards contract design, the LIMC and the client (client's family) are not linked to each other contractually. A German household in this posting model (adapted to the specificities of the live-in market) requests a domestic service brokered via the Germany agency that cooperates with a foreign agency. As a result of the brokering process, the German family signs a service contract with the foreign employment agency. The migrant care worker in turn enters into an employment relationship with the foreign agency.[8] At least theoretically, German labor health and safety standards just as the minimum wage have to apply (Böning & Steffen, 2014). This, however, is very hard to enforce in the domestic sphere.

Even less regulated and more difficult to control in terms of bogus self-employment or standards for decent work and protection is the 'self-employed placement model', another frequently used contractual design.[9] This is a model where German brokering agencies, in cooperation with, for example, Polish employment agencies, place self-employed LIMCs in German households on the basis of the EU Single Market's freedom of establishment. In these cases, the agencies help to prepare service-contracts between the LIMC and the clients, with minimum employment protection standards not applying at all (Leiber et al., 2019). Self-employment in Germany is meant to be a service provision to multiple customers, and there needs to be sufficient leeway in decision-making for the service provider on how the service is executed. Otherwise, the suspicion of bogus-self-employment is raised.

In 2014, the Posted Workers Directive was complemented by the Directive 2014/67/EU on the enforcement of Directive 96/71/EC. The main goal of the 'EU enforcement directive' was 'to prevent, avoid and combat abuse and circumvention of the applicable rules by undertakings taking improper or fraudulent advantage of the freedom to provide services enshrined in the TFEU and/or of the application of Directive 96/71/EC' (preamble). Also, the revised Posted Workers Directive (Directive EU/2018/957 amending Directive 96/71/EC), designed to further the principle of 'equal pay for equal work', was adopted in 2018, and had to be implemented by July 2020.

The (revised) PWD may be considered the most important legal base for the agency sector that we are interested in. The PWD allows agencies to construct—and advertise—an allegedly legal reference framework, although in reality even an emerging 'quality segment' (see Sect. 3 below) in the market has problems in practice to implement in a fully legal way models of employment and posting. A study by the German consumer organization Stiftung Warentest (2017) states that none of the 13 selected agencies tested could provide contracts that did not violate existing legal standards. In addition, a recent court decision attracted attention, because a state labor court ruled that—despite an existing work contract based on 30 hours per week—an employer should pay the minimum wage to a

Bulgarian LIMC for actually being on call 24 hours a day, back pay worth over 40,000 Euro (judgment Landesarbeitsgericht Berlin-Brandenburg (LAG), 17.8.2020, ref. no. 21 Sa 1900/19). An appeal of this case at the Federal Labor Court is still due, but the LAG ruling has already caused great nervousness among actors in the business field. Furthermore, the minimum legal standards that apply (in particular regarding issues of on-call duty and evaluating bogus self-employment) are not always clearly defined. There is important case law, especially on self-employment, but no legal assurance that administrative actors of the legal enforcement system act in the meaning of these complex juridical frameworks. Legal uncertainty in primary EU law on minimum social rights makes jurisdiction a crucial political issue also in this field.

In addition, the national level in Germany has a complete lack of explicit regulation for this field. To date, political stakeholders in Germany have been reluctant to address the ethically sensitive issue of migrant care work. The employment of a Haushaltshilfe (household assistant) is a legal option to employ a helper in the private household of (not only) older people. This path is also being pursued by several regional welfare organizations in Germany, acting as brokers for LIMCs, while at the same time promoting 'fair' labor contracts and standards (cf. FairCare, 2014; Menebröcker in this book). However, the actual use of this model is still rather rare because families fear a complex bureaucratic procedure and try to avoid becoming an employer. '24-hour care' is, of course, not legally possible in this way. Therefore, many clients rely more on the services of brokering agencies, promoting simple, affordable and—with reference to the EU-posting model—allegedly 'legal' solutions, promising a kind of 'convenient all-in-one package'.

To conclude, on the one hand legal uncertainty and non-transparency of the complex transnational brokering model in the EU multilevel system has enabled a new business field to emerge. On the other hand, such legal uncertainties also partly endanger evolving business models, and the involved entrepreneurs seek to preserve their increasingly professionalizing firms by using different corporate and political strategies. These will be further analyzed in the following sections in which we focus more on the corporate level.

3 Decent Care (Work): Do Agencies Make a Difference?

In our qualitative material, the companies researched[10] differ greatly in how they perceive their own profession and in their corporate governance. This leads to an overall picture of a rather heterogeneous provider landscape in which pioneering companies are taking the lead in terms of the most sophisticated 'quality' measures in their business management—in comparison to those players that are more interested in easy money and *minimal effort.*[11] The business associations in the field of live-in care work, mainly composed of *pioneers* and their *followers*, are trying to initiate a fundamental market transformation by a 'quality' push and self-imposed standards. Led by ambitious business heads, they, for instance, establish codes of conduct and voluntary agreements to be signed by all member companies to level the various business styles to a common standard (BHSB, 2015; VHBP, 2017). We observed a variety of such business practices aiming at an improvement of both the corporate (and sectoral) image and the image of the parties involved, whether customers or workers. We define these efforts as a form of corporate self-regulation that in this view 'occurs when those regulated [...] design and enforce the rules themselves. The rules that govern their behavior are adopted voluntarily, either going beyond current regulatory requirements or establishing new standards in areas which government rules or standards are lacking' (Haufler, 2013, p. 8). In the literature on corporate self-regulation (see e.g. Mills, 2016 for an overview), it is contested whether the enforcement capacity of state actors is a precondition for the establishment of private regulatory schemes, or whether 'private regulatory arrangements emerge as a result of a lack of, or failure of state-based regulatory governance arrangements' (ibid., p. 44). As we will explain below, the results in our field of interest show a mixed picture.

In this section, we seek to present our findings on such self-regulatory efforts of the live-in care agency market segment. We found that the agencies interviewed by us pursue a variety of different business strategies. When analyzing what agencies do or do not do in terms of furthering decent care and decent work based on their self-description, we sorted activities into three spheres (see Fig. 6.2). These three spheres are (1)

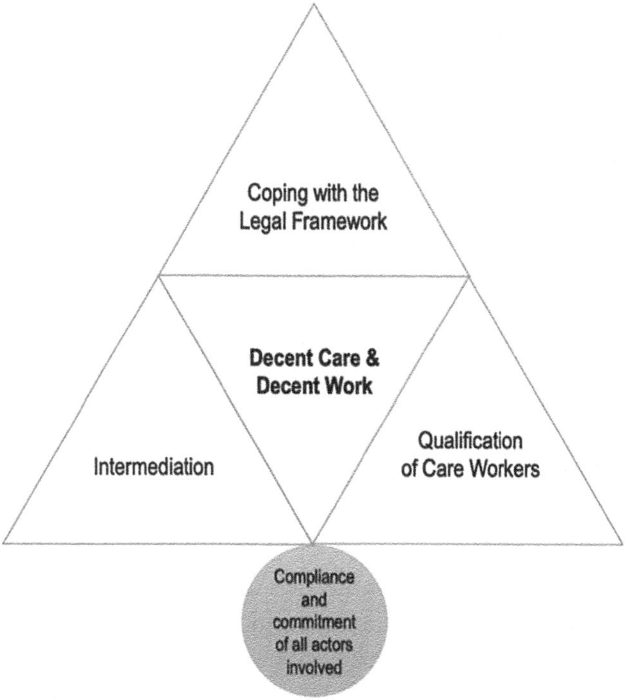

Fig. 6.2 Brokerage agencies and decent care (work) in the German live-in care market

coping with the legal framework, referring to the agencies' analysis of and coping with the given (uncertain) legal conditions (as described in Sect. 2), including aspects of contract design, minimum labor standards and social security, legal advice, and expertise; (2) qualification of care workers, that is, processes of assessment, monitoring, and further training of the workers' professional skills, as well as certification and linkage to given professional profiles; and (3) intermediation, that is, business procedures in administrative, client, and worker-related support, including monitoring, evaluation, and adaptation of working relationships, but also the professional mediation of conflicting interests.

Below this triangle, we added an element of 'instability' to illustrate the crucial point of compliance and commitment of all involved actors. Since enforcement and control by the state is practically ineffective, improving care and working conditions is strongly dependent upon this.

Coping with the Legal Framework

First, the (self-described) efforts by corporations and associations in analyzing and implementing legal conditions and legal advice provided to all actors involved require consideration. In a transnational network, the issue of coping with the legal framework covers various aspects: Which statutory rights apply? Which status do workers have according to different legal models, for example, posting or self-employment? What, for instance, does it mean when Polish Labor or Civil Law is applied in terms of rights? Corporate actors have a central position in acquiring and circulating legal knowledge and transposing it into practice. The background here is a complex legal framework, transgressing state boundaries and resulting in an intense sector network spanning Poland and Germany (cf. Fig. 6.1) that requires expert and entrepreneurial sharing of information and education (see, e.g., the fora created by a Polish think tank: Labour Mobility Initiative, 2017).

However, information asymmetries are crucial. Although German brokers demonstrate an attitude of openness and transparency, our qualitative data confirms that contractual arrangements with LIMCs are located in Poland and tend to remain beyond German jurisdiction and reach (interview, association 2). This is what we call the 'black box cooperation partner' phenomenon. There are in fact only a few German companies which are well informed about Polish contracts at all, even though there are cases where the German side works together with Polish businesses in drafting contracts with the help of experts on Polish, German, and EU law (interview, agency 1). However, as one interviewee reported, although he personally checks Polish partner contracts, he is ultimately unable to control implementation (interview, association 2; interviewee also owns an agency), for example, in terms of what is actually disbursed, or whether social insurance contributions in fact relate to all parts of the salary. This constellation is a core characteristic and weakness of the current labor market intermediation system in this sector. Some actors might not (want to) know, what's going on at the other side of the brokerage relationship, but even those actors who try have difficulties to achieve full transparency.

On the topic of contracts and the subsequent problems of transparency and judicial reliability, the political and transnational activities of business associations seem to be worth researching. In Germany, these associations introduced self-imposed 'quality standards' mandatory for all member companies (BHSB, 2015; VHBP, 2017) with the goal of both setting business standards and assuring contractual standards for this particular market segment. In February 2021 a DIN-SPEC process (i.e. a voluntary certificate setting quality standards for brokers, service providers and caregivers in the LIMC field) was initiated by dedicated agency entrepreneurs. This, however, so far covers only a very small part of the sector.

Regarding working conditions, our impression to date is that only to a limited degree improvement is provoked by such corporate interventions but—if at all—rather indirectly through a high demand for labor, which might give LIMCs certain (limited) structural negotiating powers, for example, in terms of rejecting risky clients, or very low pay (see also: interviews, agencies 1, 2, 3):

> *This means it is the market for care workers, where demand has grown enormously. This is also affected by increasing wealth in Eastern Europe. It then gets increasingly more difficult and more expensive to find new, good workers. You have to offer incentives to get the best [workers]. And this in turn results in continuously rising prices. (Interview agency 5)*[12]

Qualification of Care Workers

Decent care and decent work are equally affected by a certain level of training on the part of care workers, at best leading to an increased awareness of their own limitations, and the ability to express physical or mental strain—to the benefit of everyone involved. Central here are corporate efforts to enhance the professional and interpersonal skills of the carers themselves: Are carers well prepared for this kind of work? Have they participated in training courses? How capable are they in coping with the task of caring itself or in adequately engaging with people with cognitive disabilities? To begin with an example, we found that one German agency association was about to develop a basic training scheme based on a paragraph in the German long-term care insurance law (§45 SGB XI) that addresses training for relatives who care for dependent relatives (interview, agency 1).

However, as our data suggests, comprehensive training courses are far from being prevalent in the majority of brokerages and are being more selectively developed (e.g. Hausengel GmbH, 2018) with even the above named pilot project still under evaluation. Occasionally, pilot projects combining both mobile nursing services and live-in care brokerage have been established and may be an entrance point into a discussion of how to successfully ensure work relief schemes for live-ins in an integrated business plan (cf. interview, agency 1). Most of the preparation and training prior to taking up a post rests with the Polish cooperation partners. As our interview material reveals, recruitment approaches with varying levels of sophistication can be observed (interviews, agencies 1, 3, 5).

Some agencies, particularly the larger corporate ones, develop care-worker training courses either in the country of origin or in the destination country. One interview partner from one of the bigger companies told us their Polish partner company would check all applicants for their

> *[…] caring skills. So, first of all we check their cooking and cleaning skills which is asked about in an extended interview. We also have a hospital bed in our office and a dummy and so on. With such things, applicants can then demonstrate//how they wash hair and change diapers or so. […] and then we have a training module. I do not remember, but I think that last year we ran about 8000 training courses; from language skills to caring abilities. (Interview agency 5)*

Overall, the field of live-in care work in Germany is still far from thorough professionalization. Initial attempts toward qualification and professionalization of the care workforce can be observed but occur rarely and are not comparable to other care professions.

Intermediation

From the data we found reference to intermediation activities associated with certain aspects of decent care (work). Here, initial entrepreneurial initiatives can be observed concerning the overall process management between all parties involved. In some of the pioneering agencies (cf. Leiber et al., 2019), initial instruments and codes have been established as well as a feedback culture (customer surveys), a yearly 'quality report', the development of ethical codes (e.g. Linara GmbH, 2018; Pflege zu

Hause Küffel GmbH, 2018), and recommendations to customers when a live-in care work relationship ceases to be appropriate. Part of this intermediation of interests also concerns the practice of communicating to customers the current limitations of this kind of domestic work, sometimes in combination with the advice to use additional services in order to relieve LIMCs. Such a 'management of expectations' is important for the customer households and care workers alike (cf. Van Bochove and Zur Kleinschmiede in this book). As one head of an agency told us, his company sees itself as having a crucial 'middle position' on quality assurance in the brokerage system, between customer households and the recruitment companies in Poland:

> *We try to maintain high-quality standards for our partner companies in the sense that we select certain vacant positions [in households] but do not forward them all [to the Polish recruitment partners]. We get [customer] requests, but we only forward about 70% of them. With the remaining 30%, we do further research on them and even reject some. (Interview agency 1)*

Heads of companies, often with management backgrounds, do establish comprehensive company structures with various functional responsibilities and appropriate business processes. For instance, in-house (software) process solutions are developed and implemented to guarantee an effective and carefully targeted matching process (between the personalities and interests of both the LIMC and the household), a comprehensive monitoring and evaluation process (house visits or phone calls, statistics), as well as advisory activities on the care arrangement (e.g. assessment of care needs) and potential funding options for the long-term care scheme (interview, agency 1).

Since these companies deal with care relationships, matching tools in these settings are key (all interviews) and are becoming increasingly important in the global migratory processes of domestic workers in general. Companies use matching techniques thus as a '[...] competitive strategy. This includes offering personalized advice and expertise to employers when selecting a candidate' (Jones, 2015, p. 18).

Agencies claim that the most developed area of business activity is the close contact to households and thorough evaluations of the domestic working place by company management and employees (e.g. interview, agency 5). As the head of agency 4 told us:

> *Yes, first of all it is me who visits the customers' households to get an idea of their situation. What many others [companies] do—sitting in their office and dispatching questionnaires—does not seem productive to me. Because, I want to decide whether this person [the care worker] matches this place […].*

This head of agency and others also told us that they would routinely reject customers' requests, for instance, when 'job requirements [are] too broad-based' (interview, agency 4) or when 'too many night shifts' (interview, agency 4) are required. Similarly, agency 2 excludes households where the persons in need of care show 'aggressive behavior' or a 'strongly disturbed sleep–wake cycle.'

With regard to monitoring efforts, agency 2 phones after certain periods of time to ask customers and LIMCs about the initial days in the working relationship. Agency 5 also works with 'personal care managers' intended as contact persons for care workers in times of conflict or other work-related problems. Smaller businesses try to guarantee LIMCs around-the-clock contact option:

> *The key point is, we said, the service we offer: when a problem arises, and with these kinds of care relationships problems always arise, then people can turn to us and we can support them and clarify things. So we either settle a dispute or send the care worker back home or whatever solution emerges, right? (Interview agency 6)*

Yet, what is missing in all live-in care arrangements is the fact that, apart from the brokerage companies and the clientele having their own interests, LIMCs are usually not represented by an association or any other advocacy group they may turn to in case of maltreatment or other sensitive issues—and which exclusively focuses on their particular interests and rights.

4 Conclusion

This article seeks to shed light on a relatively new business field which has been boosted by the EU's eastern enlargement: live-in care brokerage between Germany and Poland. We have explained the international, supranational, and national regulatory environment for these businesses characterized by a lack of specific national regulation, high legal uncertainty in the EU multilevel system, and low rule enforcement in the domestic sphere. We have argued that this environment has created an additional opportunity structure—beyond the prevailing purely informal solutions—for corporate actors (corporatization) in a more formalized live-in care market. Based on a mixed methods study, we have observed that the German live-in care brokerage market may be divided into distinct segments or agency types. Certain agency types have started to position themselves as a 'quality segment'. Against the backdrop of national non-regulation, these companies and their associations have developed a variety of voluntary agreements and procedures. Our aim was to analyze the self-proclaimed 'quality' approaches of such live-in care brokerage companies, also in relation to the question of decent care (work). Do agencies as intermediaries make a difference?

We found a variety of approaches in realms we call 'coping with the legal framework', 'qualification', and 'intermediation' (see above). However, the segment of agencies we have highlighted in the article is still very small, and corporate activities differ enormously in their rigor and scope. 'Improving quality' appears to be a corporate strategy to consolidate corporate positions in the market, outrivaling competitors (from the black market) and assuring the future growth of their corporate model. Decent care (work) thus serves as a means to an end which should not be compared to the development of labor rights. In the words of the ILO: 'Self-regulation, however, cannot replace the role of national legislators and law enforcement agencies. National legislation has been seen as a means of balancing the interests of PrEA [Private Employment Agencies] with the need to protect rights of workers' (ILO, 2007, p. 2).

In this field, we see important parallels to what Lobel (2001, p. 90) found in her work on private in-home care intermediaries in the United

States and Israel: '[A]lthough market structures today provide some incentives and mechanisms for private intermediaries to enhance the conditions of careworkers, these market incentives alone are insufficient to protect careworkers from exploitation and discrimination.'

Ultimately, both the high gloss internet pages and printed brochures, particularly the suggestive power of photographs, all imply a service market that fits in seamlessly with existing domestic care services and standards (cf. Steiner et al., 2019). The hidden aspect, however, is the overriding of unresolved labor regulation and legal clarity in the field of live-in care work in Germany (with non-existent lobbying for this worker group). There are undoubtedly well-meaning elements within the self-appointed quality forays of the companies described. Equally, we see a genuine openness for political dialogue and an introduction of general standards in certain parts of the business segment, for which legal certainty (based on the Austrian model of legalized, state funded self-employed live-in care work) is central. However, this does not apply to the majority of market actors as a major share still remains in the informal shadow economy, shielded from a public view as minimum effort players (Leiber et al., 2019) or black market actors. Absent or legally uncertain state-based regulatory approaches as well as limited enforcement activities in the EU multilevel system seem to have provided the grounds for certain approaches of corporate self-regulation. But without the 'shadow of hierarchy' (Scharpf, 1997, chap. 9), that is, legislators who threaten to enact legislation unless potentially affected actors change their behavior to accommodate legislator aims, these approaches remain quite limited.

On the European level there have been initial attempts at drawing attention to this particular sector (see European Economic and Social Committee, 2016a, 2016b; European Parliament, 2016). These activities demonstrate that the question of how to deal with posted workers, domestic care workers, and also, although to a more limited degree, LIMC is on the agenda of EU institutions. Also, the revision of the Posted Workers Directive has been a noteworthy step in this respect. However, experience has shown that far-reaching steps toward positive integration into EU labor regulation standards are not easy to achieve when the interests among the member states are so diverse. And live-in care is not a phenomenon all EU member states are affected by in the same way. Therefore, it remains primarily up to national regulations to deal with the question whether and under which conditions LIMC should be part of a national care regime. In Germany, however, the growing market of live-in migrant care work has so far been politically ignored. As long as this market is not politically addressed in a more explicit way, the provider landscape of domestic care work in Germany is likely to further differentiate and grow, and certain companies will step in and develop their own rules to gain competitive advantage. In the course of this development, we have to consider, however, that formalization cannot be equated with *decency*.

Notes

1. In Germany, the country case we examine more closely in this article, the share is currently 80% (Statistisches Bundesamt, 2020).
2. Live-in care work in this article refers to foreign care workers who live for a specific period within the household of a citizen in need of care. Such care is often referred to as '24-hour care'.
3. These estimates refer to the entire workforce, meaning that these numbers include those who have returned temporarily to their home country.
4. The results presented in this article are part of the EuroAgencyCare project funded by the German–Polish Research Foundation (2016–2018). We are very grateful to Maciej Duszczyk and Kamil Matuszczyk from the University of Warsaw, who were also responsible for the Polish part of the project, for their helpful comments on this research. We also thank our student assistant Larina Kleinitz for her great help.
5. Only shortly before this book had been published, the two associations BHSB and VHBP have merged into one single association (VHBP).
6. One of the seven agencies in the German sample was not a private for-profit company but belonged to a not-for-profit welfare association. For the purpose of this article, we have omitted this case because our interest is in the role of private companies in social care provision. Additional background information was, however, gathered during interviews with political stakeholders in the EU and at the national level in Germany and Poland.

7. Employees that share a common household with those entrusted to their care and whom they rear and care for autonomously—a section that was included for the occupational profiles of foster parents in SOS Children's Villages.
8. The posting model has become very popular among live-in care brokers, in particular those in Poland. There, a domestic legal specificity (Polish Civil Law) allows recruiting and sending firms to use Umowa Zlecenie contracts that offer enough social protection to be classified as a working contract, but at the same time less social insurance contributions or workers' rights.
9. Posting-based brokering and self-employed placement are the most common models at the moment. There are other legal options, but they are rare. For an overview see Leiber et al. (2019).
10. In this section we exclusively focus on perspectives of the 'German side', that is, expert interviews with German brokerage agencies.
11. These notions stem from the analysis of Leiber et al. (2019), where a typology of German brokering agencies consisting of three types, so-called pioneers, minimum effort players, and followers, was presented.
12. All interview quotes have been translated from German to English by the authors.

References

Aulenbacher, B., Lutz, H., Schwiter, K. (Eds.). (2021). *Gute Sorge ohne gute Arbeit? Live-in-Care in Deutschland, Österreich und der Schweiz.* Beltz Juventa.

Benazha, A. (2021). Alles rechtens? Rechtliche Rahmenbedingungen der Live-in-Betreuung in Deutschland. In B. Aulenbacher, H. Lutz, & K. Schwiter (Eds.), *Gute Sorge ohne gute Arbeit? Live-in-Care in Deutschland, Österreich und der Schweiz* (pp. 46–65). Beltz Juventa.

Benazha, A., Leiblfinger, M., Prieler, V., & Steiner, J. (2021). Live-in-Care im Ländervergleich. In B. Aulenbacher, H. Lutz, & K. Schwiter (Eds.), *Gute Sorge ohne gute Arbeit? Live-in-Care in Deutschland, Österreich und der Schweiz* (pp. 20–39). Beltz Juventa.

Benner, C. (2003). Labour flexibility and regional development: The role of labour market intermediaries. *Regional Studies, 37*(6–7), 621–633.

BHSB. (2015). BHSB—Verhaltenskodex. Retrieved Mai 16, 2020, from https://www.bhsb-ev.de/verband/kodex/

BHSB. (2019). Gesundheitsminister Jens Spahn spricht sich für eine gesetzliche Neuregelung in der häuslichen Betreuung aus. Retrieved December 16, 2019, from https://www.bhsb-ev.de/2019/04/02/bhsb-vorstand-in-berlin/

Böning, M., & Steffen, M. (2014). *Migrantinnen aus Osteuropa in Privathaushalten—Problemstellungen und politische Herausforderungen*. Ver.di.

Boris, E., & Parreñas, R. S. (2010). *Intimate labors: Cultures, technologies, and the politics of care*. Stanford University Press.

Coe, N. M., Jones, K., & Ward, K. (2010). The business of temporary staffing: A developing research agenda. *Geography Compass, 4*(8), 1055–1068.

Deutscher Bundestag. (2012). Drucksache 17/8373. Antwort der Bundesregierung auf die Kleine Anfrage der Abgeordneten Klaus Ernst, Kathrin Senger-Schäfer, Jutta Krellmann, weiterer Abgeordneter und der Fraktion DIE LINKE: Arbeitnehmerrechte ausländischer Pflegehilfskräfte im grauen Pflegemarkt.

Dunger, C., & Schnell, M.W. (2018). Was ist die Framework Analysis? In M. W. Schnell, C. Schulz-Quach, & C. Dunger (Eds.), *30 Gedanken zum Tod. Die Methode der Framework Analysis* (pp. 27–39). Springer VS.

Eichhorst, W., Braga, M., Broughton, A., de Coen, A., Culot, H., Dorssemont, F., Famira-Mühlberger, U., Maarten, G., Huemer, U., Kendiza, M. J., Pedersen, J. L., & Slezak, E. (2013). The role and activities of employment agencies. Research Report No. 57 of the Institut zur Zukunft der Arbeit. IZA.

Emunds, B. (2016). *Damit es Oma gutgeht: Pflege-Ausbeutung in den eigenen vier Wänden*. Westend.

European Economic and Social Committee. (2016a). The rights of live-in carers. Public hearing of the employment, social affairs and citizenship (SOC) section on 25 April 2016. EESC.

European Economic and Social Committee. (2016b). Opinion: Fairer labour mobility within the EU, References: SOC/531 EESC-2016-00258-00-00--ac-tra. EESC.

European Parliament. (2016). Report on women domestic workers and carers in the EU. Retrieved June 3, 2020, from https://www.europarl.europa.eu/doceo/document/A-8-2016-0053_EN.html

FairCare. (2014). *FairCare—vorläufiger Abschlussbericht. Projekt für gute, gerechte und legale Beschäftigung von osteuropäischen Betreuungskräften vom 1. Januar (1. Juli) 2011 bis 31. Dezember 2013*. Diakonie Württemberg.

Farris, S. R., & Marchetti, S. (2017). From the commodification to the corporatization of care: European perspectives and debates. *Social Politics: International Studies in Gender, State & Society, 24*(2), 109–131.

Fudge, J. (2011). Global care chains, employment agencies, and the conundrum of jurisdiction: Decent work for domestic workers in Canada. *Canadian Journal of Women and the Law, 23*(1), 235–264.

Fudge, J., & Hobden, C. (2018). *Conceptualizing the role of intermediaries in formalizing domestic work*. International Labour Office.

Fudge, J., & Parrott, D. (2014). Placing Filipino caregivers in Canadian homes. Regulating transnational employment agencies in British Columbia. In K. Strauss & J. Fudge (Eds.), *Temporary work, agencies and unfree labour. Insecurity in the new world of work* (pp. 70–93). Routledge.

Haufler, V. (2013). *A public role for the private sector: Industry self-regulation in a global economy*. Carnegie Endowment.

Hausengel GmbH. (2018). Die Hausengel Akademie sorgt für einheitliche Weiterbildung. Hausengel Akademie. Retrieved Mai 16, 2020, from https://www.hausengel.de/hausengel-akademie.html

Hielscher, V., Kirchen-Peters, S., & Nock, L. (2017). *Pflege in den eigenen Vier Wänden: Zeitaufwand und Kosten. Pflegebedürftige und ihre Angehörigen geben Auskunft*. Hans-Böckler-Stiftung.

Hochschild, A. R. (2000). Global care chains and emotional surplus value. In A. Giddens & W. Hutton (Eds.), *On the edge: Living with global capitalism* (pp. 130–146). Jonathan Cape.

Ignatzi, H. (2014). *Häusliche Altenpflege zwischen Legalität und Illegalität. Dargestellt am Beispiel polnischer Arbeitskräfte in deutschen Privathaushalten*. LIT Verlag.

ILO. (2007). *Guide to private employment agencies. Regulation, monitoring and enforcement*. ILO.

ILO. (2018). *Care work and care jobs for the future of decent work*. ILO.

Jones, K. (2015). *For a fee. The business of recruiting. Bangladeshi women for domestic work in Jordan and Lebanon*. ILO.

Karakayalı, J. (2010). *Transnational haushalten: Biographische Interviews mit care workers aus Osteuropa*. VS Verlag für Sozialwissenschaften.

Krawietz, J. (2014). *Pflege grenzüberschreitend organisieren Eine Studie zur transnationalen Vermittlung von Care-Arbeit*. Mabuse.

Labour Mobility Initiative. (2017). European Labour Mobility Congress. Retrieved Mai 16, 2020, from http://www.elmc.eu/

Leiber, S., Matuszczyk, K., & Rossow, V. (2019). Private labor market intermediaries in the Europeanized live-in care market between Germany and Poland: A typology. *Zeitschrift Für Sozialreform, 65*(3), 365–392.

Leiber, S., & Rossow, V. (2016). *Self-regulation in a Europeanized 'grey market'? The role of brokering agencies in the (informal) care market between Germany and Poland.* Paper presented at the Third ISA Forum of Sociology.

Linara GmbH. (2018). Qualitätsbericht 2017. Linara. Retrieved June 8, 2017, from https://www.seniorenportal.de/ddata/13000/12947/Qualitaets-bericht.pdf

Lobel, O. (2001). Class and care: The roles of private intermediaries in the in-home care industry in the United States and Israel. *Harvard Journal of Law and Gender, 24*, 89–137.

Lutz, H. (2008). *Vom Weltmarkt in den Privathaushalt: Die neuen Dienstmädchen im Zeitalter der Globalisierung.* Barbara Budrich.

Lutz, H., & Palenga-Möllenbeck, E. (2010). Care work migration in Germany: Semi-compliance and complicity. *Social Policy and Society, 9*(3), 419–430.

Lutz, H., & Palenga-Möllenbeck, E. (2015). Care-Arbeit, Gender und Migration: Überlegungen zu einer Theorie der transnationalen Migration im Haushaltsarbeitssektor in Europa. In U. Meier-Gräwe (Ed.), *Die Arbeit des Alltags* (pp. 181–199). Springer Fachmedien.

Mills, R. W. (2016). The interaction of private and public regulatory governance: The case of association-led voluntary aviation safety programs. *Policy and Society, 35*(1), 43–55.

Peck, J. A., & Theodore, N. (2002). Temped out? Industry rhetoric, labor regulation and economic restructuring in the temporary staffing business. *Economic and Industrial Democracy, 23*(2), 143–175.

Pflege zu Hause Küffel GmbH. (2018). Ethische Grundsätze im Rahmen der 24h Betreuung. Ethische Grundsätze. Retrieved Mai 16, 2020, from https://www.pflegezuhause.info/wir-ueber-uns/ethische-grundsaetze/

Scharpf, F. W. (1997). *Games real actors can play: Actor-centered institutionalism in policy-research.* Westview Press.

Scheiwe, K., & Schwach, V. (2013). Das Arbeitszeitrecht für Hausangestellte nach Ratifizierung der ILO-Konvention 189. *NZA, 116*, 1116–1120.

Schmidt, A. E., Winkelmann, J., Rodrigues, R., & Leichsenring, K. (2015). Lessons for regulating informal markets and implications for quality assurance—The case of migrant care workers in Austria. *Ageing and Society, 36*(4), 1–23.

Schwiter, K., Berndt, C., & Schilling, L. (2014). Ein sorgender Markt: Wie transnationale Vermittlungs-agenturen für Seniorenbetreuung Im/mobilität, Ethnizität und Geschlecht in Wert setzen. *Geographische Zeitschrift, 102*(4), 212–231.

Shire, K., Heinrich, S., Imai, J., Mottweiler, H., Tünte, M., & Wang, C.-C. (2018). Private labour market intermediaries in cross-border labour markets in Europe and Asia: International norms, regional actors and patterns of cross-border labour mobility. In S. Quack, I. Schulz-Schaeffer, K. Shire, & A. Weiß (Eds.), *Transnationalisierung der Arbeit* (pp. 155–183). Springer Fachmedien.

Statistisches Bundesamt. (2020). *Pflege im Rahmen der Pflegeversicherung.* Deutschlandergebnisse 2019, Statistisches Bundesamt.

Steiner, J., Prieler, V., Leiblfinger, M., & Benazha, A. (2019). Völlig legal!? Rechtliche Rahmung und Legalitätsnarrative in der 24h-Betreuung in Deutschland, Österreich und der Schweiz. *Österreichische Zeitschrift für Soziologie, 44*(1), 1–19.

Stiftung Warentest. (2017). Trautes Heim, da will ich sein. *Test, 5*(17), 86–95.

Theobald, H., & Hampel, S. (2013). Radical institutional change and incremental transformation: Long-term care insurance in Germany. In C. Ranci & E. Pavolinie (Eds.), *Reforms in long-term care policies in Europe* (pp. 117–138). Springer.

Van Bochove, M., & Zur Kleinsmiede, D. (2019). Broadening the scope of live-in migrant care research: How care networks shape the experience of precarious work. *Health & Social Care in the Community, 28*(1), 51–59.

VHBP. (2015). Rückblick auf unseren Parlamentarischen Abend. Retrieved Mai 16, 2020, from http://www.vhbp.de/blog/posts/ruckblick-auf-unseren-parlamentarischen-abend

VHBP. (2017). Verpflichtungserklärung: VHBP-Qualitätsstandard 'Rechtliche Konformität' (Stand 5.4.2017). Retrieved Mai 16, 2020, from http://www.vhbp.de/mitglieder

Wagner, I. (2015). Arbeitnehmerentsendung in der EU: Folgen für Arbeitsmarktintegration und soziale Sicherung. *WSI-Mitteilungen, 5*, 338–344.

Williams, F. (2010). Migration and care: Themes, concepts and challenges. *Social Policy and Society, 9*(3), 385–396.

7

Caritas' Commitment to Fair and Legal Employment of Live-in Migrant Carers in Germany: A Report from Practice

Claudia Menebröcker

1 Introduction

Welfare organizations such as Caritas, the Diakonie, the Red Cross or the Malteser traditionally play an important role in the provision of social services in the German welfare state. This also applies to Germany's old age care sector. Although the country's old age care sector has increasingly been privatized, welfare organizations are still among the most important providers of long-term care. Their services range from residential care homes to day care centers to outpatient care services. Welfare organizations are recognized as nonprofit organizations and therefore entitled to public financial support. Over time, they not only have accumulated a profound knowledge in the field of old age care, but they also have become socio-political actors advocating on behalf of those at risk of poverty and exclusion. According to their self-image, welfare organizations combine 'commitment and expertise in effective forms of work and

C. Menebröcker (✉)
Caritas Association for the Archdiocese of Paderborn, Paderborn, Germany
e-mail: c.menebroecker@caritas-paderborn.de

V. Horn et al. (eds.), *The Global Old Age Care Industry*,
https://doi.org/10.1007/978-981-16-2237-3_7

organization. Due to their services to the community, they are recognized in politics and society as an important part of the welfare state' (Bundesarbeitsgemeinschaft der Freien Wohlfahrtspflege, 1992, p. 5).

Founded in 1897, Caritas is the welfare organization of the Catholic Church. In Germany, Caritas has a decentralized structure and is divided into 27 diocesan Caritas associations with 539 regional and local Caritas associations and 17 professional associations. With a total of more than 650,000 part or full-time employees, it is currently the largest employer under private law in Germany. After the fields of health and child/youth welfare, old age care is Caritas' third largest field of employment. Thus, 120,000 people or around 20 percent of all Caritas employees are currently working in the old age care sector (Caritas, 2020).

Caritas employees, particularly those working in outpatient care services, have become increasingly confronted with the phenomenon of live-in migrant carers (LIMCs) from Eastern European countries in German households. The LIMCs phenomenon has emerged to varying degrees in different European countries since the fall of communism. Driven by high unemployment rates and substantial wage differentials, many women (and a few men) from Poland and other Eastern European countries (e.g. Romania, Slovakia, Bulgaria or Croatia) opted for working as LIMCs in Germany, Austria, Switzerland and more recently also in the Netherlands (Bruquetas-Callejo, 2019; Horn et al., 2019). Estimates about the number of LIMCs working in Germany range from around 115,000–600,000 (BHSB, 2018; Hielscher et al., 2017; Ver.di, 2014). However, reliable figures are not available, since LIMCs are oftentimes employed irregularly and therefore not included in official employment registers. According to the Federal Association for Senior Citizens' Care at Home (Bundesverband häusliche Seniorenbetreuung e.V.), an association of private brokerage agencies in Germany, between 85 and 90 percent of LIMCs in Germany are employed in undeclared work (BHSB, 2018).

Private brokerage agencies position themselves as an alternative to irregular employment, offering households (allegedly) legal models of employment (e.g. under the EU posted worker directive). However, private brokerage agencies have been criticized for undermining labor rights and lacking transparency in their contracts and services (see Rossow and

Leiber in this book). By contrast, since 2009 several Caritas associations in Germany and Poland have committed to a model, which explicitly seeks to establish fair and legal employment situations for the LIMCs. Currently, 17 local Caritas associations in Germany and 1 diocesan Caritas association in Poland are involved in this model, called CariFair. They accompany about 350 families with persons in need of care and about 500-600 LIMCs employed by these families.

This chapter explains why Caritas in Germany got involved in the field of LIMCs, how the CariFair model works and how CariFair is rated by different stakeholders. Moreover, it describes Caritas' commitment at the European level and discusses the perspectives for Germany regarding long-term care for older people.

2 What Drove Caritas into the 'Agency Business'? Initial Situation for CariFair

The development of CariFair was prompted by observations by employees of outpatient care services, who increasingly reported about foreign help in the households of older people in need of long-term care. While this phenomenon is not surprising today, it was not a common model in home care 15 years ago. This situation, which was generally not legal at the time, was critical not only for the employed women and men but also for the older people in need of help and care, who had recourse to the support of foreign helpers. Less than a win-win situation, the arrangement rather represented an emergency on both sides. The LIMCs usually had no professional training as nurses and some of them did not even have experience in the field of care. This meant that those in need of (nursing) care were at risk of inadequate care, particularly if they also gave notice to the professional care service for cost-saving reasons. In addition, the employment relationships were not registered. The families and carers thus made themselves liable to prosecution for both social security and tax fraud. The question for Caritas was how they should deal with it.

Caritas works according to the mission 'See Need and Act' (Not sehen und handeln). Therefore, the organization felt obliged to improve the difficult situation for all involved. Together with Caritas Poland, four local Caritas associations in Germany developed a model to accompany these employment relationships in order to 'get them out of the grey zone'. At the beginning, Caritas took very small steps that were monitored at close intervals. First, it had to be clarified under which conditions women and men from abroad were allowed to work in Germany at all, because the free movement of workers did not yet apply to the new EU member states in Central and Eastern Europe at the beginning of the project. Qualification measures were developed and the tasks of the participating Caritas associations were coordinated. The project started with a small pilot, which was closely monitored in order to find out first of all whether and how a better situation could actually be created.

Other Caritas associations in Germany have been quite critical of the commitment of the four initiators. Some saw the deployment of LIMCs from Eastern European countries as being in competition to their Caritas outpatient care service. However, the initiators of CariFair did not share this assessment: the families hiring LIMCs simply could not afford to use an outpatient care service instead of an informal caregiver, which was why they employed the foreign care workers. Other commentators considered it unacceptable to bring women and men from Eastern Europe to Germany and thus possibly separate the families abroad. This objection is still justified. However, Caritas Poland made clear that the women and men were leaving the country anyway due to their poor financial situation—with or without Caritas. Our aim at Caritas was to provide the best possible support for employment relationships that would have existed even without us.

As independent organizations, the local Caritas associations decide for themselves which services they offer. In recent years, up to 30 local Caritas associations in Germany have participated in CariFair. However, some of them have stopped offering CariFair, for various reasons, for example, because the effort required to accompany the employment relationships was comparatively high or because too few LIMCs were found to guarantee the support of those in need of care. In addition, too few families were interested in the offer, simply because in the 'black' or 'grey' care market agencies or individuals offer much cheaper care services. Most Caritas associations—like many providers in the area of care services—now have massive personnel problems due to labor shortages in the sector. This may contribute to these associations being not able to establish CariFair as part of their offers, even if they would like to.

3 The CariFair Framework

With CariFair, Caritas associations in Germany and Poland are committed to achieve a fair balance of interests between families with older relatives in need of long-term care and the foreign staff employed by them, something that has to be negotiated again and again. On the one hand, this is about safe care for people in need of help, and on the other hand, it is about avoiding exploitation.

3.1 Employment Based on the Employer-Employee Model

Generally, three different employment models can be distinguished for LIMCs: to be hired by a German employer, to be self-employed or to be employed by a private agency from the sending state and sent to work in Germany as a posted worker. With CariFair, the employment relationship between the families of those in need of long-term care and the carers is organized as an employer-employee model. The employer of the LIMC in the CariFair concept is either the person to be cared for or their relatives. This model is also recommended by the German Labor Office and the consumer advice center (Bundesagentur für Arbeit ZAV, 2018; Verbraucherzentrale NRW, 2018, p. 31ff). The Caritas associations mediate the contact between them and the job-seeking carers. Each employment relationship is accompanied over the entire period by a Caritas coordinator.

The possibility of self-employment of the carers was ruled out in principle, as this type of employment is to be classified as bogus self-employment. This means that self-employment in this area is not possible: in everyday life, LIMCs do not have 'entrepreneurial freedom of choice' and cannot easily decide for themselves when and how they want to do which work. Instead, the LIMCs work according to instructions and depending on the needs of the person in need of care. Self-employment means that the tax and social security obligations are circumvented by the client, which is risky and possibly leads to legal problems. In addition, in principle self-employment allows unlimited working hours, which undermines the protection of caregivers. The option of cooperating with a private brokering agency that sends employees to Germany was also rejected, as the framework conditions for the employed were judged to be too opaque. In the case of agencies, it is often not clear what the costs are made up of, how much the LIMCs earn and whether and how exactly they are insured.

For families, the employer-employee-model is complicated and costly: they have to register as an employer, report their activities to the tax office, take out the necessary insurance, payroll accounting and much more. Individuals are often overwhelmed by this. For this reason, all these tasks are handled by Caritas within the framework of CariFair.

3.2 Working Conditions

Care in the context of CariFair is not 24-hour care. The aim is to work an average of 38.5 hours a week. This includes a day off during the week and regular breaks. The holiday entitlement is 36 days per year for a 6-day week. The payment of LIMCs is regulated by collective agreements. In Germany, the LIMCs pay taxes and social security contributions and are insured against accidents. In the event of illness, the LIMC is entitled to continued payment of wages for six weeks. From the perspective of Caritas, these aspects are the basis for fair and legal employment.

At CariFair, the LIMCs themselves decide how long they want to work in Germany. Many of them work on a long-term basis and only go to their home country during holidays. Oftentimes, these are employees who no longer live together with a spouse and/or children in their home country, for example, because they are divorced or their spouse is also working abroad. LIMCs whose children no longer live in their hometown also often stay in Germany for a longer period of time. Another group of LIMCs chooses a rotation model in which they come to

Germany for two or three months and then return to their home country and to their family for the same period. Some of the LIMCs also spend only a few weeks in a household in Germany and live and work in their home country for most of the year.

3.3 Accompaniment by Coordinators in Germany and Poland

One of the most important elements of CariFair is the deployment of coordinators, both in Poland and in Germany. They advise and accompany the people in need of care and their families as well as the carers. In Poland, coordinators are the first point of contact for women and men looking for work as LIMCs in a German household. The coordinators inform and advise the carers on working conditions and prepare them for work in Germany, that is, they organize language courses and basic training to accompany elderly people. The language proficiency, according to the Common European Framework of Reference for Languages, should be at least level A1.[1] However, not all LIMCs reach this level, even if they have taken a language course. If the LIMC still wants to work in Germany, she or he will be given the opportunity to try it, in coordination with the family in which she or he will be employed. An improvement in language skills often develops during the practical work. Many LIMCs have now developed quite good language skills even without a language course, simply because of the opportunity to practice in everyday life.

The basic training for accompanying older people in need of care includes basics about illnesses, including dementia, care, communication and housekeeping. Many of the carers bring along experience from their previous assignments in private households. Supplementary guidance from the local outpatient care service helps with specific questions.

The coordinators of the Caritas associations in Germany accompany the employment relationship throughout the entire period. They are the contact persons for those in need of care and their relatives as well as for the employed carers. They speak German and Polish and are committed to ensuring that the interests of both parties are safeguarded equally.

In Germany, a family looking for an LIMC would contact the Caritas coordinator. The coordinator visits the household, provides information about the employment conditions and clarifies with the family what kind of support they expect. He or she also checks the LIMC's accommodation. If both the family and the coordinator agree on the employment of an LIMC, the coordinator checks which of the available carers can be deployed. Who is currently seeking employment? Who is suitable for the specific care situation? Based on their experience, the coordinators usually have a good sense of which LIMC is likely to get along well with the person in need of care. This aspect should not be underestimated, but it is also difficult to assess on the basis of objective criteria. The coordinator assists the parties involved with the administrative requirements, such as the employment contract, insurance of the carer, registration with the tax office and the like. In addition, he or she clarifies the assignment times with both parties and discusses with them which other persons or services will be involved in the accompaniment.

Limiting working hours is the most difficult aspect, precisely because for LIMCs their workplace is also the place where they live. Solutions must also be found for holidays or if the carer falls ill. In many cases, this is only successful if relatives also contribute to the support. At the beginning of the assignment at CariFair it is therefore clarified how the free time of the caregiver can be guaranteed: Can the person in need of care remain alone on the caregiver's day off? Could the person in need of care use the outpatient care service in the mornings and evenings and visit a day care service for older adults during the day? Could the family, perhaps a neighbor, or a second carer provide the necessary support during this day? If the LIMC is absent for a longer period, the possibility of professional short-term care can also be examined. There are no generally valid solutions here. For each care situation, the coordinator agrees an individual care plan together with the family and the LIMC.

3.4 Problems in the Context of Employment

Possible conflicts and differing views on working conditions—which cannot be ruled out despite previous agreements—are dealt with together with the coordinator. Sometimes it is about seemingly small problems, which lead to arguments in everyday life, for example, if the LIMC spends too much time in the bathroom or eats too much, from the employer's point of view. Here we are dealing with conflicts that are caused by the live-in situation. They show how difficult LIMCs' working and living situations often are. Other conflicts are related to the working times of LIMCs: In spite of all the efforts, and although the observance of working hours is an important goal within the CariFair framework, the reality is that the working time limits are not observed every day. This is because the boundaries between working and leisure time are too permeable due to living together in one household.

Sometimes the coordinators reject applications for CariFair, for example, if the health situation of the care recipient is such that they cannot be accompanied by a LIMC who is not a professional nurse. This can be the case when chronic wounds are to be treated or the person concerned is on artificial respiration and the LIMC would provide care without deployment of professional nursing staff. A request for round-the-clock care by a single carer cannot be handled by CariFair either.

Sometimes there is a (health) change during the course of employment, which can no longer be handled by the LIMC alone. In this case, already existing employment relationships may be terminated if no other solution can be found.

3.5 Nursing Care by Specialists

The quality of nursing care is ensured by a binding cooperation with the outpatient care service and/or a day care service for older adults. Nursing staff can thus regularly assess whether and which further specialist nursing measures are necessary. Besides the assignment of the coordinators, the obligatory involvement of a professional care service is another important point of quality assurance, which distinguishes CariFair from the offer of many private brokerage agencies.

3.6 Financial Aspects of CariFair

The LIMCs earn about €1850 gross per month. From their salary, they pay the employee social security contribution and taxes due in Germany, leaving them—depending on the tax class—between €1100 and €1300 net per month. There are no further costs for food and accommodation. For LIMCs, CariFair advice is free of charge, and there are no costs for arranging an assignment. For the activities of the coordinators, a Caritas association currently receives €126 a month from the employer. As an employer, people in need of long-term care pay a total of around €2500 per month for the employment of a foreign carer. In comparison with other employment models, the costs for CariFair are in the middle range: for placement through private brokering agencies, families pay between €1500 and €3400 per month (Stiftung Warentest, 2017, p. 88). And even without a contract—in the case of undeclared work—the older people in need of care usually pay at least between €1200 and €1500 per month to the LIMC, according to the families.

4 Evaluation of CariFair

In 2013, the Caritas Association for the Archdiocese of Paderborn contracted the German Institute for Applied Nursing Research (Deutsches Institut für angewandte Pflegeforschung e.V., DIP) to evaluate the effects of CariFair. The aim was to record the satisfaction of all actors involved in the project, but also to identify improvement potentials and change needs. Within the scope of the study, those in need of care and their relatives, the caregivers and the coordinators from Poland and Germany were interviewed. Different methodological approaches were used, including standardized questionnaires and guided interviews. The standardized questionnaires were sent to respondents by mail; they comprised close-ended as well as open-ended questions. While the close-ended questions served to collect statements and evaluations by the respondents, the open-ended questions primarily aimed at collecting wishes and identifying problems or suggestions for the optimization of CariFair. In addition, data were gathered through telephone-based guided interviews. The standardized questionnaires were completed and returned by 103 of 396 people in need of care or their relatives, 105 of 451 caregivers and 19 of 44 coordinators. Telephone interviews were conducted with 13 LIMCs and 6 families (Isfort & Malsburg, 2014, p. 15 ff).

Overall, the survey found a high level of satisfaction with the services and offerings (Isfort & Malsburg, 2014, p. 54f). Nevertheless, DIP gave several suggestions for further development (Isfort & Malsburg, 2014, p. 56 ff.). These concerned the improvement of the language and professional competence of the caregivers, given that language courses are offered locally to varying degrees. Specialized basics are increasingly taught with the help of online courses. DIP also emphasized the great importance of coordinators, which must be secured by stable funding. Here the fee paid by the families for accompaniment has been increased. In addition to these two points, DIP called for the development of standardized information in order to make the CariFair offer transparent. For this purpose, the website www.carifair.de was developed, as well as information brochures for older people in need of care and their families and also for women and men looking for work.

Most LIMCs rate the support of bilingual coordinators as helpful. A large number of those who work through CariFair also appreciate the social security provided by health and pension insurance. In addition to carers from Poland, women and men from other Eastern European countries also turn to the coordinators in Germany in order to be able to work within the CariFair framework. They are equally accompanied in their work, but it is not always possible to ensure that they are accompanied in their native language. However, there are also carers for whom the deductions are too high and the net salary received is too low and who therefore prefer to look for irregular employment.

According to the coordinators, many people in need of care and their relatives place their trust in the 'Caritas brand'. They assess the offer as transparent and appreciate the personal support provided by the coordinators and the cooperation with supplementary services. For some of the families, however, the costs of employing an LIMC on a regular basis—which are high in comparison with an irregularly hired LIMC—are problematic. Here, stronger support from the long-term care insurance is desired.

5 Commitment of Caritas Organizations at European Level

For some years now, Caritas organizations in Eastern and Western Europe have been working together in workshops for the protection of employees and the secure care of those in need of long-term care. They have developed guidelines as a basis for their work: for a higher appreciation of care work, for a professional caregiver role in spite of their work in private households, for a representation of interests in the country of origin and the country of destination, for security and trust as a basis for a professional and respectful care relationship, and for the promotion of qualifications and better opportunities on the labor market. They try to influence European and national policies and work to strengthen civil society so that aging people can live well in all countries (Caritas Europe, 2018). Based on these guidelines, the participating national Caritas associations in Eastern and Western Europe have developed models for legal employment, qualification standards and rules for accompanying the assignments on site.

A paper of the European Economic and Social Committee (EESC) aims in a similar direction as Caritas: the EESC calls on the European Commission to recognize the work of domestic carers as a form of home care and proposes a common definition of the profession. They stress that these workers must be given the same rights as formal carers and not be excluded from relevant health and safety regulations (EESC, 2016).

At the last Caritas workshop in 2019, a paper was also adopted that makes clear that it is not enough to negotiate fair employment in the target country. The fact that people leave their home countries for employment abroad increasingly leads to care drain problems in the countries of origin. Caritas organizations are working to put the issue of care on the political agenda at the European level. The ultimate goal must be that people in all countries have access to affordable and safe care and, for this, Europe's aging society must be considered as a relevant policy field (Caritas Europa, 2019).

6 Perspectives for Germany

Undeclared work or illegal employment of LIMCs, partly under exploitative conditions, is common in German households (see above). So far, these activities have not been much regulated, but the 'market' in this field is growing steadily, as can be seen from the increase in private brokering agencies: in 2009, there were about 60 agencies and in 2017 about 266 (Stiftung Warentest, 2017, p. 88). Their offers often lack transparency. This makes the situation even more difficult for those seeking help.

Society and policymakers are aware of the problematic situation of all those involved. So far, however, there have been few attempts to put it in order. There have been various discussions at the political level in recent years, for example, with the Federal Government Commissioner for Nursing Care or the Federal Ministry of Health. However, so far, there are no discernible initiatives from these bodies to improve the situation. One could get the impression that many of those responsible in politics would rather leave the topic at rest. It is currently unclear under which conditions the topic will develop further.

We have to discuss whether the employment of migrant care workers is a good solution to mitigate the long-term care crisis—in Germany or elsewhere. However, at the moment LIMCs are simply part of our everyday life and so we as Caritas try to make these jobs as good as possible. Both the close support by a bilingual coordinator and the cooperation between the professional nursing staff and the LIMCs have proved their worth.

With the application of the current principle 'outpatient before inpatient', which assumes that care in one's own household is the best and most desirable for every person, we reach the limits of care for older people. In the long term, new ways of accompanying old people and people in need of help must be found. This could include, for example, alternative forms of housing such as multi-generational houses or shared flats for senior citizens. In these settings, round-the-clock accompaniment would be much easier to organize for several people at once. The approach of a 'caring community' could also contribute to the secure support of an aging society. 'This is not about volunteers taking over tasks from professionals or volunteers compensating the limited funds of the welfare state' (Klie, 2018, p. 21, own translation). Klie sees the division of tasks in a caring community like this:

> *Everyone contributes in his (sic) own measure to the fulfilment of the overall task: professionals guarantee professionalism, assistants reliability, relatives emotional closeness and daily support, volunteers give time, respect human rights and ensure participation. Together, the aim is to create conditions of a good life, so that people who need the support of others have the opportunity to live and realize what is important to them as human beings, even under conditions of illness, disability and need of care. No home can do this alone, no family, no outpatient care service.* (Klie, 2018, p. 21, own translation)

How to ensure that these tasks are not only taken over by a few, but that they are carried out by all in solidarity, is so far unsolved challenge.

Note

1. Level A1 refers to the first level of foreign language proficiency, also referred to as beginner level. It means, for example, that people are able to formulate very basic, everyday phrases.

References

BHSB. (2018). *Lösung für eine transparente und qualitativ hochwertige Betreuung in deutschen Privathaushalten*. BHSB.

Bruquetas-Callejo, M. (2019). Long-term care crisis in the Netherlands and migration of live-in care workers. Transnational trajectories, coping strategies and motivation mixes. *International Migration, 58*(1), 105–118.

Bundesagentur für Arbeit ZAV. (2018). Vermittlung von europäischen Haushaltshilfen. Retrieved January 24, 2020, from https://www.landratsamt-unterallgaeu.de/fileadmin/eigene_dateien/landratsamt/buergerservice/senioren/pflege_zu_hause/dokumente/Haushaltshilfen_Merkblatt.pdf

Bundesarbeitsgemeinschaft der Freien Wohlfahrtspflege (Ed.). (1992). *Freie Wohlfahrtspflege im Sozialstaat*. Bundesarbeitsgemeinschaft der Freien Wohlfahrtspflege.

Caritas. (2020). Millionenfache Hilfe—Die Caritas in Zahlen. Retrieved March 20, 2020, from https://www.caritas.de/diecaritas/wir-ueber-uns/die-caritas-in-zahlen/statistik

Caritas Europa. (2019). Final declaration on the 3rd Caritas East-West Workshop 'Fair Care Migration and Mobility in Europe'. Retrieved January 24, 2020, from https://www.caritas.eu/wordpress/wp-content/uploads/2019/11/191129-Fair-Care-final-declaration.pdf

Caritas Europe. (2018). Fair care mobility and migration in Europe. Retrieved January 24, 2020, from https://www.caritas.eu/fair-care-mobility-and-migration-in-europe/

EESC. (2016). The rights of live-in care workers. Official Journal of the European Union C 487, 28.12.2016. Retrieved January 24, 2020, from https://eur-lex.europa.eu/legal-content/EN/TXT/?uri=uriserv%3AOJ.C_.2016.487.01.0007.01.ENG&toc=OJ%3AC%3A2016%3A487%3ATOC

Hielscher, V., Kirchen-Peters, S., & Nock, I. (2017). *Pflege in den eigenen vier Wänden: Zeitaufwand und Kosten. Pflegebedürftige und ihre Angehörigen geben Auskunft*. Hans-Böckler-Stiftung.

Horn, V., Schweppe, C., Böcker, A., & Bruquetas-Callejo, M. (2019). Live-in migrant care worker arrangements in Germany and the Netherlands: Motivations and justifications in family decision-making. *International Journal of Ageing and Later Life, 13*(2), 83–113.

Isfort, M., & Malsburg, A. (2014). *Evaluation des Projektes Heraus aus der Grauzone—Qualitätsgesicherter Einsatz polnischer Haushaltshilfen in deutschen Familien mit pflegebedürftigen Angehörigen*. DIP.

Klie, T. (2018). Keiner sorgt für sich allein. *Publik-Forum EXTRA, 11*(2018), 20–22.

Stiftung Warentest. (2017). Pflege zu Hause: Die besten Vermittler für Betreuungskräfte aus Osteuropa. *test* 05/2017, 86–95.

Verbraucherzentrale, N. R. W. (2018). *Ausländische Haushalts- und Betreuungskräfte in Privathaushalten*. Verbraucherzentrale NRW.

Ver.di. (2014). *Migrantinnen aus Osteuropa in Privathaushalten. Problemstellungen und politische Herausforderungen*. Ver.di.

8

Pragmatic Inattention and Win-Win Narratives: How Finnish Eldercare Managers Make Sense of Foreign-Born Care Workers' Structural Disadvantage

Antero Olakivi and Sirpa Wrede

1 Introduction

In Northern Europe, care providers increasingly employ foreign-born workers, but mainly in lower-level jobs that are relatively underpaid, overburdened and unattractive to native-born workers, especially in care for older persons (e.g. Dahle & Seeberg, 2013; Wrede & Näre, 2013; Cangiano & Walsh, 2014). Consequently, critical social science has expressed concerns over emerging 'ethnic hierarchies' in the labor market and the field of care work (Laurén & Wrede, 2008). In these hierarchies, racialized minorities, foreign-born persons and other people in generally disadvantaged labor market positions (Hellgren, 2015) are overrepresented, if not blatantly 'exploited' (Näre, 2013), in societally undervalued and burdensome employment settings. Misra et al. (2006, p. 318), for example, write about the 'international division of care work that places

A. Olakivi (✉) • S. Wrede
Centre of Excellence in Research on Ageing and Care, University of Helsinki, Helsinki, Finland
e-mail: antero.olakivi@helsinki.fi; sirpa.wrede@helsinki.fi

V. Horn et al. (eds.), *The Global Old Age Care Industry*,
https://doi.org/10.1007/978-981-16-2237-3_8

the [heaviest] burden for care on the least powerful (immigrant women workers)'. The emerging 'migrant division of care labour' (Näre, 2013) is fueled by the politics of economic austerity, due to which the attractiveness of eldercare jobs has decreased in Northern Europe. These developments add to the pre-existing recruitment problems deriving from the persisting low level of recognition of eldercare work in Northern Europe and globally. In circumstances where austerity has become institutionalized as the central mode of public governance, organizations have applied new recruitment and employment practices to manage alarming labor shortages (e.g. Cangiano & Walsh, 2014; Wrede et al., 2021). This chapter focuses on how local eldercare managers discursively engage with the emerging workforce constellations.

Recent research has examined the complex ways in which employers in formal (e.g. Näre, 2013; Cleland Silva, 2018) and informal (e.g. Gavanas, 2013; Gallo & Scrinzi, 2016) care participate in the prevailing, austerity-driven recruitment and employment practices. However, this research on employers and their representatives, such as managers, who are responsible for recruitment and contracting, remains relatively limited. According to Cangiano and Walsh (2014, p. 373), 'theory on hiring migrant [care] workers, particularly in terms of a sociological understanding, is not well developed'. According to Gallo and Scrinzi (2016, p. 367), 'employers' role in driving the demand for foreign-born [care] workers is considered a key issue in social sciences, [but] it remains largely under-researched and under-theorised'. This chapter targets this gap by shedding light on the discursive means through which local eldercare managers legitimate the prevailing recruitment and employment practices as solutions to the alarming labor shortages in Helsinki, Finland.

The chapter makes two contributions to research on the prevailing recruitment and employment practices in care work. First, most research on such practices has focused on care employers who are actively involved in hiring workers across national borders and from the 'international' labor market (Adhikari & Melia, 2015; Hellgren, 2015; Cleland Silva, 2018). Such employers are directly involved in the global care industry, contributing to global care chains, possible care drains in underprivileged countries and the potential exploitation of global inequalities (Yeates,

2009; Wrede, 2010; Lutz, 2018). The research focus on international recruitment is understandable, yet it may maintain the unintended impression that only international recruitment exploits structural inequalities and marginalized actors' disadvantaged labor market positions. 'Domestic' recruitment, however, can also exploit racialized and migration-based labor market structures and inequalities, albeit in the national setting. In this chapter, we focus on eldercare managers who recruit foreign-born workers mainly from the domestic labor market. However, we also question the categorical distinction between 'international' and 'domestic' recruitment and argue that actual recruitment practices can occasionally be more complex and ambiguous than this dichotomy implies.

Second, research on recruitment for eldercare and beyond (e.g. Waldinger & Lichter, 2003; McGovern, 2007; Anderson & Ruhs, 2010) often 'focuses more on why migrant workers are recruited than on how' (Cangiano & Walsh, 2014, p. 373). In this chapter, we shift the research focus to the study of how—with a specific focus on moral justifications in and around the prevailing recruitment practices (also Weicht, 2011; Nordberg, 2016; Torres, 2017). In short, we ask: How can eldercare managers discursively counter the potential criticism according to which the prevailing recruitment practices exploit actors in structurally disadvantaged labor market positions and, among them, foreign-born workers? We pay particular attention to justification practices that mitigate the importance of structural improvements in the position of foreign-born workers and, ultimately, in austerity-driven eldercare work in general. Before we further introduce our analytical approach and analysis, we briefly discuss the context of Helsinki.

2 Migrant Divisions of (Care) Labor in Helsinki

In global comparison, Finland relies heavily on formal, professional and publicly funded eldercare provision, despite international policy trends of privatization and marketization also have a growing impact in Finland. In

Finland, the 1990s constituted a turn toward austerity policies as governments sought to control public expenditure for care services. Policies targeted eldercare in particular, with reforms outlined toward increasing emphasis on home care, together with aims to increase cost-efficiency in care work (Henriksson & Wrede, 2012; Karsio & Anttonen, 2013). Further in line with international trends, the Finnish provision of eldercare increasingly relies on foreign-born workers. This reliance is particularly prevalent in the capital city Helsinki (Statistics Finland, 2019), and especially at the lowest occupational levels. In 2016, the proportion of foreign-born workers in Helsinki's total workforce was 11%, whereas the proportion of foreign-born workers as care assistants was 24%, practical nurses 15%, registered nurses 10% and head and ward nurses 1%. Foreign-born workers were also overrepresented in social care (17%) compared to health care (11%). In Finland, care work in social services is commonly regarded as less prestigious and more burdensome than care work in health services (Henriksson & Wrede, 2012).

Among foreign-born care workers in Finland, the most common countries of birth are Estonia, the Philippines, Russia and Somalia (Koivuniemi, 2012; Aalto et al., 2013; Ailasmaa 2015). In international comparison, the volume of recruitment from the Philippines has remained relatively limited in Finland (Näre & Nordberg, 2016). By 2018, the two recruitment agencies that dominate this business had recruited circa 300 educated Filipino nurses to Finland, catering for their enlistment and placing, but not for the certification of their nursing degrees. Consequently, Filipino nurses primarily work in positions below their formal qualifications, mostly as carers in private care homes (Vartiainen, 2019). The recruitment for City of Helsinki both targets foreign-born people living in Finland and pursues recruitment abroad. Within the city organization, international recruitment is delegated to the human resources department, which functions independently from care organizations. A particular focus lies on workforce from nearby regions, as the City continuously advertises job opportunities in the care sector in Estonia. The backgrounds of foreign-born care workers, and their recruitment routes in particular, are not well-documented in Finland. Most foreign-born care workers appear to have migrated to Finland primarily for reasons other than to specifically work in care (e.g. as refugees, along other humanitarian pathways or for family reasons), but the available statistics do not offer detailed information on this matter. The lack of official documentation is not surprising, since the Finnish political authorities have also remained somewhat disengaged with questions concerning the ethics of recruitment and, in particular, of international recruitment, leaving considerations over ethics, including the potential care drains in the countries of origin, to individual employers (Vaittinen, 2017).

How should one make sense of the recruitment and employment tendencies in Helsinki? Media presentations in Finland (Nordberg, 2016), as in many other countries (e.g. Torres, 2017), often present the prevailing recruitment and employment tendencies as serving multiple interests in tandem. In these narratives, the increasing employment of foreign-born workers in eldercare is a solution to the aging Finnish society's need for eldercare workforce, but it also caters for the needs of foreign-born workers and international job seekers to gain access to employment and income. According to critics, however, the lack of workforce in the Finnish eldercare sector is not only caused by the aging population but also by the low wages and unappealing working conditions in eldercare, caused by the prevailing austerity policies (Wrede, 2010; Wrede & Näre, 2013; Näre & Nordberg, 2016). Currently, the number of Finnish-born eldercare workers planning to quit their jobs is soaring, mainly due to the under-resourcing and psychophysical overload in eldercare work (Olakivi et al., 2021). In Finland, the percentage of eldercare workers who were seriously considering leaving their jobs grew from 26% in 2005 to 41% in 2015 (Kröger et al., 2018). Critics thus argue that the prevailing recruitment and employment tendencies might serve Finnish employers and economically oriented policymakers better than they serve foreign-born workers (Näre, 2013; Wrede, 2010). Instead of fostering the quality of care work, employers and policymakers hold on to austerity practices and focus on workforce who, due to unequal labor market practices, including racial discrimination in various employment settings (Wrede et al., 2010; Nieminen, 2011; Larja et al., 2012), have limited options in other fields of employment.

3 Care Managers' Discursive-Affective Practice

How can eldercare managers negotiate their moral position in relation to the above developments, pressures, interests and critiques? To a degree, eldercare managers are accountable for their actions in and around recruitment (Shutes & Walsh, 2012). To some stakeholders, eldercare managers may have to justify how the prevailing recruitment and employment tendencies serve Finnish-born care recipients (despite, e.g., public concerns over foreign-born workers' language skills, see e.g. Dahle & Seeberg, 2013). To others, critical questions relate to the potential exploitation of foreign-born workers and their disadvantaged labor market positions. In all cases, eldercare managers' justifications may require a degree of pragmatic inattention to critical counter-narratives. The empirical analysis that follows pays particular attention to such practices of inattention, and the consequences of these practices for eldercare organizations. Narratives can for example empower and disempower organizational reforms and social transformation programs. As Silverman (2017, p. 156) points out: 'Wars are fought and elections are won and lost in part on the basis of compelling narratives. The role of historians and social scientists is to unpick how stories are constructed whether in political speeches or qualitative interviews.'

Our empirical analysis builds on qualitative interviews of 14 eldercare managers (e.g. head nurses, ward managers and home care supervisors) working in one public sector nursing home and in six different districts of public home care in Helsinki. The interviews were conducted by three researchers in 2011–2013 in a project entitled, 'Shaping of Occupational Subjectivities of Migrant Care Workers: A Multi-sited Analysis of Glocalising Elderly Care', funded by the Academy of Finland (#251239). The interviewed managers were all women and had different responsibilities in and around recruitment. In the nursing home, the head nurses were formally responsible for enlisting permanent staff, including writing job advertisements and interviewing applicants. In home care, this formal responsibility belonged to a specific recruitment department. Ward managers and home care supervisors were responsible for recruiting temporary workers and fixed-term substitutes. In practice, however, they also participated in the recruitment of permanent staff, including interviewing applicants, alongside head nurses and the recruitment department. In general, all the participants highlighted human resource management as an important aspect of their work (Olakivi & Niska, 2017). In the nursing home, about one-third of the care workers were foreign-born. The home care units were selected for the study on the condition that they employed more foreign-born workers than home care units on average in Helsinki.

The interviews contained several themes, ranging from ethnic diversity and the recruitment of foreign-born workers to the future of eldercare in Finland (for a detailed description of the data, see Olakivi, 2018). The interviews were audio-recorded and transcribed verbatim. The extracts provided below have been translated by the authors with some details omitted to protect the participants' anonymity (in line with what we promised to the participants). The analysis is based on the original recordings.

We analyze the interviews as fragments of discursive-affective practice (Wetherell, 2013). Through the interviews, we thus examine the discursive practices through which eldercare managers can affectively 'experience themselves as good, valuable, worthwhile people' (Lamont & Swidler, 2014, p. 159) despite the complex pressures they face in their work. In the interviews, we allowed the participants to choose their own words, tell the narratives they found affectively appealing and—if they felt a need—justify their actions and inactions in and around recruitment.

Interview narratives are not individual inventions but they draw on socially circulating discursive resources such as culturally recognizable identities and subject positions, impressions of personal agency, vocabularies of motive and knowledge about socio-political structures (see Clarke et al., 2009; also Lamont & Swidler, 2014). Although our analysis is based on research interviews, eldercare managers can—or have to—draw upon similar resources in other contexts, including interactions with care workers and care recipients. The power of discursive resources derives from their presence in public discussions. Therefore, critical attention must target the institutions that mobilize these discourses, including the media (e.g. Weicht, 2011; Nordberg, 2016; Torres, 2017). Eldercare managers are merely the end users of publicly circulating discursive resources.

Our analysis was led by the following question: How can eldercare managers discursively counter the potential criticism according to which the prevailing recruitment and employment practices exploit the structurally disadvantaged labor market position of foreign-born workers? To answer this question, we analyzed all the interview extracts in which the managers discussed the recruitment of foreign-born workers. Below, we examine three qualitatively different discursive-affective practices: the mitigation of personal agency, the construction of win-win situations and the distinction between domestic and international recruitment. The analysis sheds light on the organizational background of these discursive practices, as well as their potential, unintended consequences for eldercare organizations.

4 Findings

4.1 No Agency, No Blame, No Prestige?

Social science literature occasionally criticizes eldercare managers for strategically and intentionally recruiting workers in structurally disadvantaged labor market positions and, among them, foreign-born workers (Näre, 2013; Cangiano & Walsh, 2014). Discursively, however, eldercare managers have several ways to counter this criticism (also Olakivi, 2020). In our interviews, eldercare managers were—often quite easily—able to point out structural limitations to their recruitment activities and thus construct themselves as non-agents (see Clarke et al., 2009) in recruitment, as in the following extract.

Extract 1

INTERVIEWER: Have you made conscious efforts to recruit people with foreign backgrounds here?[1]

PARTICIPANT12 [home care supervisor]: Not consciously, but like … A normal search [for applicants], and then we of course interview a couple of competent, suitable [applicants]. Often the case in home care is that there is only one or two applicants ((laughs)).

In the above, the participant highlights the structural limits to her recruitment activities, namely the limited number of job applicants. The participant does not explicate the reasons for this limit but assumes it to be self-explanatory. It is plausible to interpret that she relies on the common knowledge according to which the deteriorating working conditions in home care have decreased the sector's attractiveness among care workers in Finland (Henriksson & Wrede, 2012; Kröger et al., 2018).

In the above line of discursive-affective practice, eldercare managers are not personally responsible if foreign-born workers, or other persons in disadvantaged labor market positions, end up being overrepresented in care jobs that otherwise attract few applicants. This impression of non-agency, however, has a negative side effect. It constructs an image of recruitment as relatively routine and mechanical work that does not require high levels of specialization, skills or expertise from eldercare managers; anyone can do recruitment. To eldercare managers, this image may not be affectively appealing. In order to construct themselves as important and prestigious actors, eldercare managers may have to present recruitment as work that involves selection and decisions, as in the following extract.

Extract 2

PARTICIPANT9 [home care supervisor]: I always trust first impressions, I don't know if that's a good or bad thing. That's why I always want to see the person myself. Even if someone else, after interviewing the person, says how good they are and so on, I always feel that 'fine, but it was not my decision'. I want to influence it myself.

When eldercare managers construct a degree of personal agency in recruitment, they must ensure that this agency serves legitimate interests: those of care recipients and, ideally, also care workers. The following section examines the discursive-affective practices through which eldercare managers can, at least momentarily, construct legitimate and affectively appealing versions of their personal agency.

4.2 Good Workers, Nice Jobs

Eldercare managers have a variety of discursive resources with which they can legitimate their personal engagement with the prevailing recruitment practices. As many studies have noted (e.g. Näre, 2013; Shutes & Walsh, 2012; Walsh & Shutes, 2013), eldercare managers can draw on ethnic and racialized stereotypes and construct foreign-born workers as ethnic others who are both skillful care workers and inherently interested in eldercare work. In the following extract, a ward manager highlights her personal agency in this way.

Extract 3

INTERVIEWER: Have you made conscious efforts to recruit people with foreign backgrounds?

PARTICIPANT5 [ward manager]: Well, we have indeed had these migrants who've been coming here. ... I've taken them as apprenticeship students. We've had NN1 and NN2 [anonymized] as apprenticeship students. [...] But I mean yes, yes I have, quite consciously.

I: Why?

P5: Because in my opinion, they have the kind of qualities ((laughs)) that they respect older people, and it's innate in them. I mean, the innate conduct they have. I have had a lot of [foreign-born workers], temps and all, and they've made very good impressions.

I: Yeah, and ... So, in your view it relates to ... Are there differences in respect to different origins, or do you see that it's in general ... Although they come from different cultures?

P5: Yes I believe so. Although they come from different cultures, they still have stronger ties to older people. The respect. It's probably in the mother's milk, I think. It's totally different to us Finns, for example.

In the interviews, the managers constructed African and Asian workers in particular as actors who have 'totally different', culturally inherited traits and abilities—such as respect for older people. In the above, foreign-born employees are, moreover, constructed as inherently interested in eldercare work. Therefore, their overrepresentation in the eldercare sector can hardly be seen as a tendency that exploits unequal labor market structures. The above line of discursive-affective practice can thus, intendedly or not, 'morally legitimatize and/or neglect new hierarchies and inequalities in working life' (Nordberg, 2016, p. 104).

Ethnic and racial stereotypes are not the only discursive resources through which eldercare managers can justify their personal agency in the prevailing recruitment practices and, ultimately, the emerging migrant divisions of care labor. In the following extract, a ward manager draws on another discursive resource to construct the prevailing recruitment and employment tendencies as win-win tendencies. The extract continues a broader discussion (not shown here) over inadequate economic resourcing in eldercare in Finland, and over the difficulties the participant has faced in finding skilled and motivated workers to enlist. In the extract, the participant constructs foreign-born workers as members of an international community of care workers who are ethically committed to serving older people and who can serve Finnish care recipients irrespective of their language skills. The cited ward manager, for example, employs temporary workers with whom, according to her (not shown here), she does not share any common language.

Extract 4

INTERVIEWER: Do you think the importance of language has been overemphasized perhaps [in discussions about eldercare work]?

PARTICIPANT7 [ward manager]: Somehow I feel that it is sometimes. Because, at least here, we never receive feedback on this from the relatives [of the residents]. It's never been brought up as a problem in the discussions. The biggest problem that comes up is the lack of time [for care]. [...] Not the language. Of course, I do understand that it's [important]. [...] But if things continue the way they are, then I think we should loosen up the [language] requirements and not be so awfully strict in respect to certain applicants. We should somehow collect all resources together and see that whatever country we're from, it's certain that, as care workers, we all serve the same purpose and work here for the older people.

In the above, the ward manager draws on a variety of discursive resources to justify the recruitment of foreign-born workers for eldercare even if their language skills do not match the formal requirements. First, the need to recruit foreign-born workers results from a structural necessity. Second, the prevailing (or suggested) recruitment practices serve all actors in tandem: 'We all serve the same purpose'. Compared to the first justification that highlights structural limits, the latter justification seems more inspiring, optimistic and affectively engaging. In the above, the manager also positions herself as a care worker (see 'as care workers, we all …'), further mitigating potential hierarchies and conflicts of interest.

Whereas Extract 3 constructed foreign-born workers as members of their national and ethnic groups (and consequently skilled and motivated workers), Extract 4 constructs foreign-born workers as members of an international community of care workers (and consequently skilled and motivated workers). Eldercare managers thus have different ways of presenting the increasing employment of foreign-born workers in eldercare as a legitimate, if not desirable tendency. One can evaluate these discursive-affective practices in different ways. On the one hand, eldercare managers can, through these practices, justify the prevailing recruitment practices to audiences who, due to racist or xenophobic sentiments, would prioritize the recruitment of Finnish-born (or white) workers. On the other hand, these discursive-affective practices limit the subject positions available for foreign-born workers in eldercare organizations. To support these narratives, foreign-born workers must appear as inherently motivated to work in eldercare, regardless of the sectors' increasingly austere working conditions. In this respect, the above lines of discursive-affective practice, in Extracts 3 and 4, are relatively similar: they both turn attention away from structural disadvantages.

Previous research highlights how employers' or managers' stereotypical and essentialist constructions of foreign-born workers as ethnic others can contribute to migrant divisions of care labor (Carter, 2000; Näre, 2013; Cangiano & Walsh, 2014). These constructions reduce the opportunities for foreign-born workers to have their formal merits (e.g. education) recognized and, consequently, to achieve higher professional ranks. Ethnic othering constructs foreign-born workers as fit to work close to the recipients of care, but not in managerial positions (e.g. Dahle & Seeberg, 2013; also Nordberg, 2016). These concerns must be taken seriously. At the same time, these concerns may underestimate the *diversity* of discursive resources with which eldercare managers—and other stakeholders—can justify, and turn attention away from, the emerging migrant divisions of labor and the potential exploitation of actors in structurally disadvantaged labor market positions.

According to Burr (2003, p. 137), 'an important part of warranting one's actions, of making them appear reasonable and justifiable, is having the ability to present oneself in different ways according to the demands of the moment'. Evidently, eldercare managers occupy a morally precarious position (Shutes & Walsh, 2012; Olakivi, 2020): they have to justify their activities, including the activities they cannot change, in front of different audiences (e.g. care recipients, care workers, other eldercare managers and municipal authorities). Ethnic stereotypes may work as viable resources in some contexts of justification, whereas adopting a 'colorblind' discourse and constructing foreign-born workers as members of an international community of care workers may satisfy other audiences. Both lines of discourse can mitigate the importance of systemic and structural improvements in foreign-born workers' labor market position—and in the conditions of eldercare in general. Improvements are not needed, since all actors already receive what they authentically need: the care recipients receive good care and the care workers receive jobs that align with their interests. This alignment of interests is achieved, in part, by assuming that employment in eldercare serves foreign-born workers without the need to improve care workers' wages or working conditions. This assumption is explicitly invoked in the following extract.

Extract 5

INTERVIEWER: Then what about immigrants?

PARTICIPANT2 [head nurse]: Immigrants. The way I see it, there's an awful lack of workforce in this field of work, and it's only getting worse. Soon it'll be a real [issue]. Where are we getting employees from? And in my view, people with migrant backgrounds are one of those resources, and many [of them] are interested. This is a really nice workplace. A nice workplace. And it's not the nationality. It's a nice workplace [also] for a Finn. There are many opportunities to develop and … No opportunities to get a really good salary, but opportunities to proceed [career-wise] and develop the work and so on.

In the above, the head nurse first employs the resource of non-agency in the recruitment of foreign-born workers: foreign-born workers are an indispensable resource in the context of an alarming lack of workforce. The possibility of tackling the lack of workforce by, for example, improving care workers' wages or working conditions is not discussed in the extract. After highlighting the lack of alternatives, however, the participant constructs a different version of the role of foreign-born workers in eldercare. In this version, employment in eldercare serves the interests of the recruited workers: many of the workers are interested and can be offered good jobs, or at least jobs with career prospects. To make this narrative sound convincing, the manager ignores the overrepresentation of foreign-born workers in lower-level care occupations and, thus, their evident career stagnation. Evidently, the manager lacks the discursive resources with which she could construct contemporary, austerity-driven eldercare work as an ideal job (for anybody), but for a foreign-born jobseeker, perhaps one with fewer better alternatives or little interest in a good salary (in the Finnish standards), it is a decent job (also Torres, 2017). The head nurse carefully manages the impression of a win-win situation, though not without a discursive struggle. As this struggle demonstrates, it is not always easy for eldercare managers to suppress potential counter-narratives.

4.3 Against International Recruitment: Or Out of Sight, Out of Mind?

In addition to the construction of non-agency and win-win situations, the interviewed managers had another discursive resource for justifying their participation in the prevailing employment tendencies, regardless of how these tendencies might exploit structural inequalities in the labor market. This resource constructed a sharp distinction between the 'domestic' and 'international' recruitment of foreign-born workers. Evidently, constructing international recruitment (from poorer countries) as a tendency that violates the interests of Finnish-born workers, care recipients and occasionally also the recruited workers is a pragmatic resource for local eldercare managers (also Olakivi, 2020). With the help of this resource, eldercare managers can construct their own, domestic recruitment practice as an unproblematic and respectable, if not desirable, alternative. In the following extract, Participant 7, who previously called for collecting 'all resources together' (Extract 4), argues against international recruitment.

Extract 6

INTERVIEWER: What do you think about the international recruitment of nurses or care workers, what if the municipality would start doing it? Some private care companies do it already, but what would you think if the City of Helsinki would start doing it?

PARTICIPANT7 [ward manager]: I'm against it, as a matter of fact. They do recruit a lot of nurses from the Philippines, and it's great that they can be offered jobs. But in my opinion, we should start to think. ... There are so many nurses without a job in Finland. We have so much educated care personnel, hiring them should be the priority, in my opinion. It cannot be the solution that the kind of cheaper workforce is recruited from the Philippines because they're willing to work for these low wages.

In the above, the ward manager conceives problems in the employment practices that, instead of improving wages in care work, seek to recruit workers in disadvantaged labor market positions. The ward manager, however, relates these problems to 'international' recruitment, not 'domestic' recruitment. The manager thus externalizes the problems of austerity-driven recruitment to other recruitment agents. In reality, the employer of the interviewed ward manager, the City of Helsinki, already recruits care workers from, for example, Estonia to work in eldercare in Helsinki. In the interviews, the managers—and the interviewers—occasionally demonstrated a degree of inattention, amnesia or ignorance in respect to these practices of recruitment.

Extract 7

INTERVIEWER: What about direct recruitment from abroad, what do you think of that, when the private sector and also the Helsinki University Hospital has recruited directly? Would it be possible to do that here at some point?

PARTICIPANT1 [nursing home director]: If you have to go there, then. But I'd rather start with having a vacancy, and then the person who wants it applies for it. There is always also the thing that … Sometimes with these foreigners, when they come from different cultures, their jobs may have been completely different there. When we really have informal work outfits and so on here, then maybe for some it has been a surprise, if they have been looking for a position as a clinical nurse, then this does not quite match [with their expectations]. But maybe that's in the future then. But I wouldn't go to recruit anyone at least now ((laughs)). But then again, for our work, the human resources department [of the City of Helsinki] is constantly going to, for example, Estonia to recruit. But it is recruiting for everywhere [i.e. all jobs in the municipality], not just for us. But I don't know if anyone has actually arrived here that way. But they go there to talk [about job opportunities] and share contacts.

In the above, the nursing home director argues against international recruitment that does not always serve the recruited workers. Instead of mentioning how the austere working conditions or limited career prospects in Finnish eldercare organizations might come as a surprise to the recruited workers, the director invokes another problem that, one might argue, turns attention away from the structurally disadvantaged position of the recruited workers: the recruited workers might expect the work to be more 'clinical'. Only after arguing against international recruitment does the director seem to remember that the City of Helsinki in fact already does recruit internationally. At this point, the director demonstrates a degree of ignorance regarding her employees' routes to work: 'I don't know if anyone has actually arrived here that way'.

To managers who want to maintain a moral distance from 'international' recruitment, a degree of *ignorance* regarding the routes through which their employees have arrived can be a pragmatic strategy (see McGoey, 2012). In this way, eldercare managers can maintain a sharp distinction between 'domestic' and 'international' recruitment and disengage themselves from the problems they associate with the latter. In Helsinki, this moral disengagement is institutionally managed by the City, which has delegated international recruitment to the human resources department, which continuously advertises job opportunities in the care sector in Estonia. The human resources department works separately from the care organizations. Consequently, eldercare managers can assume they are mainly participating in the affectively appealing practices of domestic recruitment, without always knowing their applicants' actual connections to Helsinki and Finland.

5 Discussion and Conclusions

This chapter has shown how eldercare managers at the floor-level of care organizations can discursively and affectively participate in the prevailing recruitment and employment practices that, according to critics, have a tendency to exploit racialized and migration-based labor market inequalities and, ultimately, leave the root causes of labor shortages in eldercare intact. Our analysis demonstrates how the recruitment of foreign-born workers to eldercare employment, also from 'domestic' labor markets, is a morally complex practice to eldercare managers—that is, a practice that requires rationalizations and justifications in relation to various (real or imaginable) counter-arguments (in research interviews, but assumedly also in other contexts) (also Olakivi, 2020). Our analysis has further illuminated the different ways in which eldercare managers can discursively disengage themselves from the structural disadvantage of foreign-born workers and, ultimately, from the prevailing austerity policies.

Research has often focused on practices of ethnic stereotyping and othering that can justify the emerging divisions of labor in which foreign-born workers are overrepresented in lower occupational echelons and practically excluded from leadership roles. In this chapter, however, we have demonstrated how ethnic othering is not the only practice through which eldercare managers—and other stakeholders—can justify the emerging divisions of labor and, ultimately, downplay criticism against the austerity policies that exploit such divisions. Eldercare managers can end up justifying the overrepresentation of foreign-born workers in lower-level eldercare jobs also by discursively constructing an international community of care workers that shares an ethical commitment to eldercare work but has little interest in high (or even decent) wages or improvements in working conditions.

Eldercare managers thus have different discursive resources with which they can—often unintendedly—make the emerging divisions of care labor seem tolerable, if not affectively appealing—at least to themselves.

Foreign-born workers, of course, do not always agree with their managers' win-win narratives (Olakivi, 2017). Foreign-born workers may, for instance, express discontent over their wages, working conditions, career prospects or, simply, structurally disadvantaged labor market positions. Such expressions of discontent conflict with their managers' idealized narratives. This conflict, we argue, partly explains why foreign-born workers' expressions of discontent are so easily suppressed in eldercare organizations' internal communications (Olakivi, 2017, 2018). Expressions of discontent do not sit well with the win-win narratives that many eldercare managers find personally appealing.

Eldercare managers can also highlight their non-agency vis-à-vis the prevailing, austerity-driven practices of recruitment and employment. Regarding macro-societal divisions of labor, constructions of non-agency are easily available to floor-level eldercare managers who are hardly responsible for the exclusion of foreign-born workers from higher-status jobs. Regarding recruitment, however, impressions of non-agency may lack affective appeal in a context that values eldercare managers' personal ability to make a difference in their environment (also Gallo & Scrinzi, 2016). It seems that eldercare managers are under institutional pressure to present themselves as agents who can serve Finnish-born care recipients (Olakivi & Niska, 2017), but also foreign-born staff-members. This institutional pressure, we argue, partly explains why these managers are so eager to construct foreign-born workers as inherently interested in eldercare work. These constructions enable eldercare managers themselves to adopt affectively appealing positions as 'agents for' (Niska 2015) several actors in tandem. To maintain their affectively appealing positions, however, eldercare managers may have to suppress counter-narratives that call for improvements in the position of foreign-born workers.

Local eldercare managers' criticism of 'international' recruitment can further help them justify their own, 'domestic' activities in recruitment. A sharp distinction between 'domestic' and 'international' recruitment helps eldercare managers maintain affectively appealing impressions of themselves as actors who simply recruit employees from the unproblematic labor market in Finland, in contrast to 'other' actors who are engaged with more problematic, and potentially exploitative, practices of international recruitment. The upkeep of this moral boundary may require, however, a degree of inattention from eldercare managers to their employees' routes to Finland. Sometimes these routes can be more complex than implied in the sharp distinction between 'domestically' and 'internationally' recruited workers. Using a distinct human resource department as a proxy agent who is in charge of international recruitment, while eldercare managers maintain their moral distance from these recruitment practices, can be a pragmatic strategy for local eldercare managers.

Finally, we wish to highlight that our aim in this chapter has not been to problematize the inclusion of foreign-born workers in the care workforce, not even in the most underpaid, overburdened and austerity-driven eldercare jobs. The quality of these jobs does not get any better if eldercare managers stop recruiting foreign-born workers. Instead, our chapter calls for critical attention toward societally circulating, optimistic and affectively appealing win-win narratives that turn collective attention away from the austerity-driven devaluation of eldercare jobs *and* from the structural disadvantage of foreign-born workers. These win-win narratives, often mobilized in the media (e.g. Weicht, 2011; Nordberg, 2016; Torres, 2017), can disempower alternative narratives—circulating in the margins of eldercare organizations and the media—that call for structural improvements in the quality of eldercare jobs and the position of foreign-born workers in and beyond the care industry.

Note

1. In the extracts, brackets signal removed passages or added clarifications and underlining indicates that a word was emphasized by the speaker.

References

Aalto, A. M., Elovainio, M., Heponiemi, T., Hietapakka, L., Kuusisto, H., & Lämsä, R. (2013). *Ulkomaalaistaustaiset lääkärit ja hoitajat suomalaisessa terveydenhuollossa: haasteet ja mahdollisuudet*. National Institute for Health and Welfare.

Adhikari, R., & Melia, K. M. (2015). The (mis)management of migrant nurses in the UK: A sociological study. *Journal of Nursing Management, 23*(3), 359–367.

Ailasmaa, R. (2015). *Terveys- ja sosiaalipalveluiden henkilöstön kansainvälinen liikkuvuus 2012*. National Institute for Health and Welfare.
Anderson, B., & Ruhs, M. (2010). Migrant workers: Who needs them? A framework for the analysis of staff shortages, immigration and public policy. In M. Ruhs & B. Anderson (Eds.), *Who needs migrant workers? Labour shortages, immigration, and public policy* (pp. 15–52). Oxford University Press.
Burr, V. (2003). *Social constructionism*. Routledge.
Cangiano, A., & Walsh, K. (2014). Recruitment processes and immigration regulations: The disjointed pathways to employing migrant carers in ageing societies. *Work, Employment & Society, 28*(3), 372–389.
Carter, J. (2000). New public management and equal opportunities in the NHS. *Critical Social Policy, 20*(1), 61–83.
Clarke, C. A., Brown, A. D., & Hailey, V. H. (2009). Working identities? Antagonistic discursive resources and managerial identity. *Human Relations, 62*(3), 323–352.
Cleland Silva, T. (2018). *Transnational management and globalised workers: Nurses beyond human resources*. Routledge.
Dahle, R., & Seeberg, M. L. (2013). Does she speak norwegian? *Nordic Journal of Migration Research, 3*(2), 82–90.
Gallo, E., & Scrinzi, F. (2016). Outsourcing elderly care to migrant workers: The impact of gender and class on the experience of male employers. *Sociology, 50*(2), 366–388.
Gavanas, A. (2013). Elderly care puzzles in Stockholm: Strategies on formal and informal markets. *Nordic Journal of Migration Research, 3*(2), 63–71.
Hellgren, Z. (2015). Markets, regimes, and the role of stakeholders: Explaining precariousness of migrant domestic/care workers in different institutional frameworks. *Social Politics: International Studies in Gender, State & Society, 22*(2), 220–241.
Henriksson, L., & Wrede, S. (2012). The making of medico-managerial care work culture in public home care for the elderly. Perspectives on care at home for older people. In C. Ceci, K. Björnsdóttir, & M. E. Purkis (Eds.), *Homecare: International and Comparative Perspectives* (pp. 171–185). Routledge.
Karsio, O., & Anttonen, A. (2013). Marketisation of eldercare in Finland: Legal frames, outsourcing practices and the rapid growth of for-profit services. In G. Meagher & M. Szebehely (Eds.), *Marketisation in nordic eldercare: A research report on legislation, oversight, extent and consequences* (pp. 85–126). Stockholm University.

Koivuniemi, S. (2012). *Maahanmuuttajataustainen koulutettu hoitohenkilöstö sosiaali- ja terveydenhuollon työyhteisöissä*. Tehy ry.

Kröger, T., Van Aerschot, L., & Puthenparambil, J. M. (2018). *Hoivatyö muutoksessa. Suomalainen vanhustyö pohjoismaisessa vertailussa*. University of Jyväskylä.

Lamont, M., & Swidler, A. (2014). Methodological pluralism and the possibilities and limits of interviewing. *Qualitative Sociology, 37*(2), 153–171.

Larja, L., Warius, J., Sundbäck, L., Liebkind, K., Kandolin, I., & Jasinskaja-Lahti, I. (2012). *Discrimination in the Finnish labor market: An overview and a field experiment on recruitment*. Ministry of Employment and the Economy.

Laurén, J., & Wrede, S. (2008). Immigrants in care work: Ethnic hierarchies and work distribution. *Finnish Journal of Ethnicity & Migration, 3*(3), 20–31.

Lutz, H. (2018). Care migration: The connectivity between care chains, care circulation and transnational social inequality. *Current Sociology, 66*(4), 577–589.

McGoey, L. (2012). The logic of strategic ignorance. *The British Journal of Sociology, 63*(3), 533–576.

McGovern, P. (2007). Immigration, labour markets and employment relations: Problems and prospects. *British Journal of Industrial Relations, 45*(2), 217–235.

Misra, J., Woodring, J., & Merz, S. N. (2006). The globalization of care work: Neoliberal economic restructuring and migration policy. *Globalizations, 3*(3), 317–332.

Nordberg, C. (2016). Outsourcing equality: Migrant care worker imaginary in Finnish media. *Nordic Journal of Working Life Studies, 6*(3), 101–118.

Näre, L. (2013). Ideal workers and suspects. Employers' politics of recognition and the migrant division of labour in Finland. *Nordic Journal of Migration Research, 3*(2), 72–81.

Näre, L., & Nordberg, C. (2016). Neoliberal post colonialism in the Media: Constructing Filipino nurse subjects in Finland. *European Journal of Cultural Studies, 19*(1), 16–32.

Nieminen, S. (2011). *Kuulumisen politiikkaa: Maahanmuuttajasairaanhoitajat, ammattikuntaan sisäänpääsy ja toimijuuden ehdot*. University of Tampere.

Niska, M. (2015). *Who is serving whom? An agency-for perspective on enterprise promotion in rural Finland*. University of Helsinki.

Olakivi, A. (2017). Unmasking the enterprising nurse: migrant care workers and the discursive mobilisation of productive professionals. *Sociology of health & illness, 39*(3), 428–442.

Olakivi, A. (2018). *The relational construction of occupational agency. Performing professional and enterprising selves in diversifying care work*. University of Helsinki.
Olakivi, A. (2020). The problematic recruitment of migrant labor: A relational perspective on the agency of care work managers. *Current Sociology, 68*(3), 333–352.
Olakivi, A., & Niska, M. (2017). Rethinking managerialism in professional work: From competing logics to overlapping discourses. *Journal of Professions and Organization, 4*(1), 20–35.
Olakivi, A., Van Aerschot, L., Puthenparambil, J.M., Kröger, T. (2021). Ylikuormitusta, lähijohtajan tuen puutetta vai vääränlaisia tehtäviä: miksi yhä useammat vanhustyöntekijät harkitsevat työnsä lopettamista? *Yhteiskuntapolitiikka, 86*(2), 141–154.
Shutes, I., & Walsh, K. (2012). Negotiating user preferences, discrimination, and demand for migrant labour in long-term care. *Social Politics, 19*(1), 78–104.
Silverman, D. (2017). How was it for you? The Interview Society and the irresistible rise of the (poorly analyzed) interview. *Qualitative Research, 17*(2), 144–158.
Statistics Finland. (2019). Labour statistics on migrant and foreign-born employees available on request. http://www.stat.fi.
Torres, S. (2017). Fobbing care work unto 'the other'—what daily press reporting shows. *Sociologisk forskning, 54*(4), 319–322.
Vaittinen, T. (2017). *The global biopolitical economy of needs: Transnational entanglements between ageing Finland and the global nurse reserve of the Philippines*. University of Tampere.
Vartiainen, P. (2019). *Filippiiniläisten sairaanhoitajien polut Suomeen: Tutkimus oppimisesta ja työyhteisöintegraatiosta kansainvälisen rekrytoinnin kontekstissa*. University of Tampere.
Waldinger, R., & Lichter, M. I. (2003). *How the other half works: Immigration and the social organization of labor*. University of California Press.
Walsh, K., & Shutes, I. (2013). Care relationships, quality of care and migrant workers caring for older people. *Ageing & Society, 33*(3), 393–420.
Weicht, B. (2011). Embodying the ideal carer: The Austrian discourse on migrant carers. *International Journal of Ageing and Later Life, 5*(2), 17–52.
Wetherell, M. (2013). Affect and discourse–What's the problem? From affect as excess to affective/discursive practice. *Subjectivity, 6*(4), 349–368.

Wrede, S. (2010). Nursing: Globalization of a female-gendered profession. In E. Kuhlmann & E. Annandale (Eds.), *Palgrave handbook of gender and healthcare* (pp. 437–453). Palgrave Macmillan.

Wrede, S., Nordberg, C., & Forsander, A. (2010). Etnistyvä eriarvoisuus ja keinot kohti tasa-arvoa. In S. Wrede & C. Nordberg (Eds.), *Vieraita työssä: työelämän etnistyvä eriarvoisuus* (pp. 274–283). Gaudeamus.

Wrede, S., & Näre, L. (2013). Glocalising care in the nordic countries: An introduction to the special issue. *Nordic Journal of Migration Research, 3*(2), 57–62.

Wrede, S. Näre, L., Olakivi, A., Nordberg, C. (2021, accepted for publication). Neoliberal 'flexibility' and the discursive incorporation of migrant labour in public eldercare in Finland. In N. Piper & C. Mora (Eds.), *The palgrave handbook of gender and migration: Global perspectives.* : Palgrave Macmillan.

Yeates, N. (2009). *Globalizing care economies and migrant workers: Explorations in global care chains.* Palgrave Macmillan.

9

Social Dynamics and Concepts of Good Care Affecting the Interaction Between Established Employees and Newcomers in a German Nursing Home

Anita Ham

1 Introduction

The shortage of skilled care personnel is a major problem in many countries, especially those with aging populations (World Health Organization, 2020). As one of the most 'super-aged' societies in the world[1], Germany must prepare for nearly one-third of its population to be aged 65 or older by 2050 (Federal Statistical Office, 2016). The aging of the population will exacerbate the already existing shortage of skilled health care workers and long-term care personnel. The shortfall in nurses has been a long-standing issue, with the last decade witnessing health care vacancies filled by nurses recruited from Korea, Russia, Poland, and other countries (Bonin et al., 2015). In 2015, 20% of all health care institutions recruited

A. Ham (✉)
Institute of Public Health and Primary Care, Leiden University Medical Center, Leiden, The Netherlands

The Hague University of Applied Sciences, The Hague, The Netherlands
e-mail: a.ham@hhs.nl

V. Horn et al. (eds.), *The Global Old Age Care Industry*,
https://doi.org/10.1007/978-981-16-2237-3_9

caregivers from Eastern Europe (Federal Statistical Office, 2015; Rada, 2016). The German federal employment agency and the Deutsche Gesellschaft für Internationale Zusammenarbeit (GIZ) set up the 'Triple Win' project to bring qualified nurses from Serbia, Bosnia and Herzegovina, the Philippines, and Tunisia to Germany (GIZ, 2015). In addition, between 100,000 and 300,000 Eastern European migrants are delivering home-based care (Hanewinkel, 2015; Kniejska, 2015), mainly as live-in migrant care workers looking after the elderly in private households (Horn et al., 2019). Despite this expansion in labor sources, the shortage in caring staff continues to worsen. Especially in nursing homes there are many unfilled vacancies (Bundesministerium für Gesundheit, 2018).

In 2015, as a result of the so-called refugee crisis[2] that challenged many European countries, almost 500,000 asylum applications were submitted in Germany (ECRE, 2020). Among the applicants were professionals who had obtained nursing degrees in their countries of origin, giving rise to hope in the German health care industry that these nurses would help to reduce the shortage of staff in various health care facilities.

Although certain nursing practices are similar globally, what is considered 'good' in nursing is also socially, culturally, and politically constructed and influenced by the national context and organizational rules and regulations. Many immigrant nurses experience challenges in their host country, like discrimination, culture shock, and language difficulties (Kingma, 2018; Xiao et al., 2014; Zhou et al., 2011). According to a study by Ulusoy and Schablon (2020), in German nursing homes discrimination of immigrant health care workers occurs in the form of xenophobia and refusal of care by residents.

This current study is part of a larger research project. The first study focused on the workforce integration[3] of first-generation immigrant health care workers with refugee backgrounds in a nursing home in the Netherlands (Ham, 2019) and analyzed how these workers collaborated with established staff to enact good care (Ham, 2020). This chapter is based on a study conducted in a German nursing home for elderly dementia patients in 2017. The aim was to contribute to the evaluation of a project for recently arrived asylum seekers and refugees by exploring the social dynamics between the established employees and the newcomers and examining how concepts and practices of good care affected their interaction.

This research involved shadowing 17 refugees and asylum seekers recruited by the St. Johannes Health Care Institution (JHCI, a pseudonym) which offered them a three-month basic nursing training program. Simultaneously, all established staff members received a training to become interculturally competent workers—that is, to develop a collegial attitude of respect, regardless of gender, nationality, ethnic origin, religion, disability, age, or sexual orientation. The theoretical framework of the study consisted of Elias and Scotson's (1965) established–outsider configuration and Bourdieu's concept of habitus (1990) to examine the social dynamics between the established health care workers and the newcomers in this German nursing home.

2 Understanding Group Processes

Elias and Scotson (1965) studied the effects of the arrival of newcomers on an established community in a small town. The residents upheld strong social networks that had made them a close-knit community, a Gemeinschaft, based on shared norms and habits. The newcomers, on the other hand, came from various places, so they could not rely on a shared social network as a source of power. Elias and Scotson (1965) argued that structural inequalities are based on power relations between two groups. Together they form a configuration of interdependent groups, the 'established' and the 'outsiders.' Both groups were striving for control of the same power sources. However, the established group was more powerful, due to their duration of residence, their strong internal cohesion, and their monopolization of power positions in the community. The authors showed how, in such a constellation, certain mechanisms reinforce the boundaries between groups and induce the development of collective images of 'us versus them' (Elias & Scotson, 1965). Mechanisms like stigmatization and gossip can be seen as powerful weapons of the dominant group (established) to exclude the other group (outsiders). Elias (1994) claimed that such a social dynamic could be explained by differences in length of residence and power resources rather than by racial, ethnic, class, or gender differences. He argued that this division of power between groups should be central in any analysis of social divisions.

Bourdieu (1990) maintained that when people are part of a particular field, such as the field of education or the field of health care, they develop a particular 'habitus,' which can be defined as the ingrained habits, skills, and dispositions of individuals, which manifest themselves in a routinized system of rules and foster social cohesion. People cultivate the specialized knowledge of their specific field, thereby accumulating social capital. By developing 'a feel for the game' (Bourdieu, 1990, p. 66), they gain a practical sense of how to act in a given situation in a field. Therefore, people who are familiarized with their specific field are likely to share perceptions and dispositions that translate into similar practices and representations.

3 Methodology

3.1 Setting and Sample

The JHCI is a German nursing home that provides care to psychogeriatric, medically ill, and terminally ill elderly residents. From 2017 to 2019, the institution carried out a project called Asylbewerber und Flüchtlinge als Potenzialträger [AFP] in Deutschland, ein Modellprojekt im Sozialwesen (Asylum Seekers and Refugees as High Potentials in Germany, a Model Project in Social Affairs). The aim of the project was to provide refugees and asylum seekers the opportunity to explore if they would like to be trained as nurses, assistants, or helpers in nursing homes. With help from the job and municipal center, churches, and nearby asylum centers they recruited a total of 69 newcomers to participate in the project, originating from Syria, Afghanistan, Iraq, Iran, and Kurdistan; African countries (e.g., Burundi, Nigeria, Togo, and Eritrea); and Eastern European nations (e.g., Macedonia and Kosovo). At the time of this study, most of them were awaiting the decision on their asylum application and had not yet received a residence permit.

The JHCI trained five groups (between 12 and 18 participants per group) as nurse aide for dementia residents [*Alltagsbegleiter/in für Menschen mit Demenz*], in five successive AFP projects, each lasting three months. Two months were devoted to vocational education (328 hours), and one month was dedicated to internships (150 hours). The two-month AFP program was an orientation of a vocational nursing training for care workers. The program dealt with basic knowledge and skills of psychogeriatric care, German language skills, health prevention, the German health care system, and team collaboration. Art and culture projects, such as theater and role play, were also part of the training. The last month was for internships in the psychogeriatric wards. These three months were followed by an evaluation in which the newcomer and the project manager ascertained the best fit for the newcomers' future opportunities. For example, the newcomer could begin with an education for nurse assistant (*Krankenpflegerassistent/in*), helper (*Altenpflegerhelfer/in*), or choose to eventually continue with higher education to become a registered nurse (*KrankenPfleger/in* or *Altenpfleger/in*).[4] A few newcomers were offered a voluntary social year [*Freiwilliges Soziales Jahr*] in the nursing home—for example, to improve their German language skills. The project manager wanted to help the newcomers to achieve their objectives, but she obviously hoped that they would choose a nursing career and help alleviate the labor shortage in the (elderly) health care sector.

3.2 Data Collection

For this research, data were collected during the third AFP project, which lasted from July to September 2018. To ensure validity, triangulation was performed through participant observation (shadowing), interviews, and focus group discussions. The focus was on the newcomers and established health workers who had supporting or supervisory roles during the training or internship. Seventeen newcomers were observed during the classes and three (two with a nursing background and one with nursing experience) were shadowed during the internships. Participant observation was conducted two days a week during classes, meetings, and actual work (200 hours).

Further, 24 in-depth interviews were conducted with six newcomers and seven established staff members, six management employees, and five teachers. On the basis of the observations, the interviews focused on (1) the AFP project as well as whether and how (mutual) learning and adaptation were taking place, (2) challenges and barriers encountered between the established health professionals and the newcomers, and (3) respondents' notions of good care. The interviews took place in the nursing home and lasted between 45 minutes and 60 minutes. All sessions were audio-recorded and transcribed verbatim in German.

Two focus group discussions were conducted, one with six newcomers and one with six established employees. Based on the observations and interviews, we discussed (1) the AFP project and processes, (2) collaboration between the established health professionals and the newcomers, and (3) their notions of good care. The focus group discussions were also held in the nursing home and lasted between 90 and 120 minutes. The sessions were audio-recorded and transcribed verbatim in German.

3.3 Ethical Considerations

The study was approved by the Medical Ethical Review Committee of the Leiden University Medical Center in the Netherlands (code P 16.087). The researcher, a Dutch medical anthropologist with a background in nursing as well as the author of this study, collected the data. The researcher was asked by the AFP project manager (a German anthropologist and nurse) to monitor one of the AFP projects in the nursing home through an ethnographic study for a project evaluation. Written information was distributed among the participants, including informed consent forms, letters with additional details, and guarantees of confidentiality. During meetings in the nursing home, the aim and method of the research were explained to the prospective participants in German and English, after which informed consent was obtained from them. The nursing home and all the participants were assigned pseudonyms.

3.4 Data Analysis

The data consisted of field notes made during the observations and transcripts of the interviews and focus group discussions. The notes and transcripts were read repeatedly to familiarize with the data. Inspired by a post-positivist grounded theory coding approach, data were analyzed using ATLAS.ti. Information saturation occurred after repeated reviews of the sampling, data collection, and data analysis. With a combination of inductive and deductive analyses, a deeper understanding of the social dynamics between the established staff members and the newcomers was gained. We utilized the notions of 'established' and 'outsiders' (Elias & Scotson, 1965) and 'habitus' (Bourdieu, 1990). To check validity, the main findings were presented to the participants (member checking), who confirmed that the results reflected their views, feelings, and experiences.

4 Results

As previously described, 17 newcomers, all refugees and asylum seekers originating from Iraq, Syria, Afghanistan, Togo, Uganda, and Azerbaijan, participated in this third part of the AFP project. The newcomers (8 females and 9 males aged between 20 and 55 years old) had diverse educational backgrounds, ranging from engineering and accountancy to pedagogy and philosophy, and their work experience ranged from 0 to 30 years. Although the nursing home wanted to recruit newcomers with a nursing background, only two of the 17 newcomers had obtained a nursing diploma in their home country, while one had gained nursing experience during the Syrian war. Moreover, it must be noted that the foreign nursing diplomas were not recognized by German authorities.

The established staff consisted of mostly female carers of German, Polish, Yugoslavian, Bulgarian, Russian, Congolese, and Turkish descent aged between 18 and 65 years. They worked in the wards as nurses [*Altenpfleger/in*], nurse assistants [*Pflegerassistent/in*], or as care assistants [*Altenpflegerhelfer/in*]. Their work experience varied from 2 to 30 years.

In order to ensure successful collaboration between the established employees and the newcomers, the former had completed a short training program on intercultural competence[5] (1.5 hours each week for eight weeks). The aims of this training program were to make diversity visible, stimulate dialogue, underline the significance of diversity in all its dimensions, and raise awareness in dealing with the similarities and differences between people.

4.1 Organizational Constraints

When the newcomers began their internships, the managers matched each newcomer with an established 'buddy' to work the same hours and shifts. The established workers had not been asked if they wanted to become a buddy. Some of them wondered why they were obliged to participate, but were afraid to inquire further. Some established employees took the newcomers on board as part of their team and supported them and taught them the rules and regulations of the wards. Others, including nurses who were assigned as buddy, were unwilling to work with the newcomers given their already burdensome workload and hectic time schedules. They neglected the new colleagues entirely; they merely divided the tasks, after which the newcomer was left to work on his or her own. Belo, a German-Congolese man who had been working in the nursing home for a few years, realized that more employees were required, but he also had mixed feelings about the project: 'It takes so much energy and what I received in return was nothing. You do not gain benefits from your own investments because after three months they leave and a new group comes in, which means you need to start all over again.'

The established staff were convinced that if the management had involved them in the decision-making process, better matches could have been made between the buddies and the new arrivals and the strategy would have been beneficial for all. The newcomers were aware of the shortage of staff and felt a responsibility to help both their colleagues and the residents. However, a few of them encountered organizational constraints as well. For example, there were (single) parents who had no one to look after their small children when they had to work, which made it difficult to handle the irregular work shifts.

During the research period, in addition to the already stressful situation, the nursing home suffered a severe outbreak of scabies.[6] The institution attempted to control this through a series of measures, such as administering drugs, washing and disinfecting clothing and beddings, and sanitizing the entire premises of the home. The caregivers were also required to wear protective clothing (i.e., disposable long-sleeved gowns and gloves). All these were rather time-consuming tasks and exacerbated the already heavy workload.

Samsu, a newcomer from Iraq, was scared to transmit this contagious infection to his wife and children. He became anxious, especially when he listened to established Polish worker Jana who suffered from scabies on and off. Jana's skin reddened and itched, preventing her from sleeping at night. Her scabies had been treated many times to no avail. Samsu feared that the mites might hide in his beddings or in the furniture or carpeting of his home.

The wearing of protective clothing on top of regular apparel was rather uncomfortable as it was a very hot summer, with temperatures occasionally rising above 40 °C. The established employees were told that during the recent renovation of the building, no money was left for climate control. Xena, one of the care assistants, was displeased that the management had not consulted them, the care workers, during the renovation; after all, they were the ones working in the wards. Others thought that the people who were in charge of the institution did not care about them.

Thus, organizational constraints and institutional structures (rules and regulations) and the feelings of the established workers that their views and perceptions and/or the issues they face were not heard and appreciated by the management of the institution (along with other factors and mechanisms) hindered intergroup relationships and interactions, as will become evident in the following sections.

4.2 Different Concepts and Practices of Good Care and Language Difficulties

Both groups—the established employees and the newcomers—were outspoken regarding their conception of what good care is. In certain respects, they upheld the same norms. For example, most of them believed that providing good care implies that you invest in the relationship and are friendly to and supportive of the dementia residents. However, the established staff regarded their own norms as superior to those of the newcomers. They considered themselves better workers because they took their responsibilities very seriously, exhibited intrinsic motivation, and worked hard. Karlotta, a German nurse, wished that the newcomers would take their everyday tasks and responsibilities more seriously. She claimed that tardiness can be overlooked when it happens once only, not more. The presumptions of the established staff regarding the supposedly inappropriate behavior of the newcomers negatively affected their relationship.

In a certain sense, the comments of the established staff were contradictory. On the one hand, they perceived the newcomers as requiring considerable mentoring, given the latter's unfamiliarity with institutional norms, standards, and routines. They were afraid that the newcomers would fail to live up to their concepts of good care. They asserted that this work requires professionalism, skills, and expertise. On the other hand, they seemed to assume that daily care was an easy task. A number of them, for example, were sure that showing the newcomers how to work on day one would enable them to know what to do by day two. After all, this had also been the way they had been instructed themselves when they started to accomplish caring duties in the wards. They became irritated when repeated explanations of tasks could not drive the newcomers to actually 'see' the work. They complained that this inability only added to their onerous workload. As one of the established employees exclaimed: 'some don't even see the work, even when we are telling them what to do! I really don't understand, as the work is the same when they wake up and get dressed!'

Some of the established workers averred that good caring comes straight from the heart and that this was what was lacking in some of the newcomers. They suggested that many of the newcomers were not intrinsically motivated, as showed by their calling in sick frequently or arriving late at work.

Newcomer Mohammed, on the other hand, said that he did not always know what was expected from him when he was working independently, because he was not familiar with the work and dementia residents. This showed when I attended a morning shift:

It is 08:00, I walk upstairs to the first floor. In the ward, I see Mohammed, a nurse from Iraq with whom I am working today, standing in front of Karlotta: 'Look,' says Karlotta screaming, pointing her fingers in the direction of Mr. Zweiger ..., 'Look,' she says again, 'at Mr. Zweiger ...! How is this possible?' In the kitchen, Mr. Zweiger, sitting on a trolley, wearing white shirt and underpants, a urine pad could be seen through his white slip. He was barefoot and eating a sandwich. 'I specifically asked you to wash him, put on his shirt, his sweater, his trousers, his socks, and his shoes, then to shave him, comb his hair, and then take him to the dining room for his breakfast,' Karlotta shouts. 'And now look at Mr. Zweiger,' she continues, pointing at him again, 'now he is there, half naked, eating his breakfast!'

The established employees doubted whether unskilled workers from outside Europe with no nursing qualifications would be able to learn to care for German dementia patients in such a short time. They were skeptical about whether a two-month orientation was sufficient. However, a few of the established workers with immigrant backgrounds themselves acknowledged that the new arrivals reminded them of their own situation when they were starting out in Germany and had to adapt to a different (health) system.

Kabiba, a Turkish established health care worker, for instance understood why some of the newcomers refused certain tasks. She linked this to their different culture: 'You know what it is; men from these countries do not do these kinds of things, like bathing, washing, or helping other people go to the toilet. These are females' tasks. So, we cannot force them to do these now that they are here.' Moreover, she did not expect that those men, with their entirely different educational backgrounds, would choose a nursing career in Germany.

Conversely, the newcomers had their own reasons for refusing tasks. Instead of Kabiba's cultural explanation, Brev for instance gave a more personal reason. Trained as an engineer in Syria, he particularly struggled with helping residents to the toilet or changing diapers because it made him nauseous. He said that he could not even change his children's diapers without throwing up, let alone those of 'strangers.' He recounted one of his work experiences in the following manner:

> *We came to Mrs. Doldermann's room, and we found her naked on the floor. It was horrible; I just could not stand it. Not only was she completely naked, but she had smeared faeces all over everything, like the chair, the floor, her hair, her face, her breasts; everything was covered in excrement. I had to leave the room right away. I could not stand the smell. … I was almost vomiting. I could not clean it either; I just cannot do this kind of work.*

Other newcomers justified their refusal by referring to what they had been taught during the two-month AFP educational program. Although they were informed in class what they could expect, practice proved to be quite different. Neram, a Syrian male worker, for instance refused to perform tasks outside his expertise, such as assisting in giving a resident a shower in a slippery bathroom, because he worried about who would be held responsible if one of his patients fell. He explained that during the training, they were instructed to refrain from doing tasks that were beyond their comfort zone. In practice, however, it appeared that a 'don't ask, don't tell' approach dominated. The newcomers said that they attempted to work according to the routines, but the expectations of the established staff were unclear for them.

Some of the newcomers argued that their behavior was more caring than that of the established staff. Bibi, a newcomer from Iraq, insisted that she behaved more respectfully toward the elderly residents than her established counterparts. She observed that the morning washing of the residents was done rapidly, as though the care workers were always in a rush. This observation was confirmed by Junis, a newcomer from

Afghanistan who stated that she at least had patience and could give the elderly people what they needed:

Everything in the nursing home has to be done very quickly, you know, and the staff here are impatient with the residents. For example, they just put the food in front of them, and five minutes later, they collect it again, without even looking if their patients have finished it. No, I am not like that, you know, I just sit with them quietly and respectfully and help them finish their food.

During their internship, the newcomers came to realize that integration into the German nursing workforce required acclimatizing to the political, social, and cultural values that underpin the German health care system. A feel for the game, that is, an intuitive understanding of how things work was needed to acquire this new habitus. Some of the newcomers doubted that they would be able and/or willing to adapt to how things work in German nursing homes. Mohammed seems to be a case in point: even though he had gained substantial experience as a nurse in his home country, he doubted whether a nursing career in Germany would be his best option. Not only was the work different from that in his home country, but one was also required to acquire a good command of the German language:

In Iraq, nurses measure blood pressure, give injections and medicine, and they do not wash patients like they here in Germany. Although I have worked for 13 years as a nurse in Iraq, I don't think I can work in a nursing home in Germany! Here a nurse has mainly a social function, so for me, it would be better to find a job as a bus driver because then the language would not matter.

Mara, from Iraq, was once reduced to tears after having received the news that she had failed the B2 German state exam[7] for the second time. Leyla, a Syrian newcomer, also pointed out the importance of language skills in providing good care, admitting that she found German very difficult to understand, particularly the narratives of the elderly and all the nuances. She also felt excluded and frustrated when a few of her established colleagues spoke only Polish, which impeded her from improving her ability to communicate in German. The native German workers acknowledged that some of the Polish employees in their team were also not that good at speaking the German language but, as Sebastian said, 'they are just part of our team.'

Thus, factors like the different concepts and practices of good care and language difficulties influenced the social dynamics between the established staff and the newcomers. Feelings of superiority among the established staff were not only based on their familiarity with the formal rules and procedures but also on their professional habitus, which they claimed to be based on a better notion of good care (e.g., caring from the heart, working hard, and being intrinsically motivated). The newcomers did not have this habitus, as most of them did not have a background in nursing. Practice turned out to be rather different from what they had been taught in class. Some newcomers refused tasks, others thought of themselves as better than the established staff. The newcomers experienced language difficulties and realized that they needed a 'feel for the game' to acquire a new habitus in this nursing field.

4.3 Resentment and Pain from Indifference

When the established health care staff were sitting together in the nursing staff room, which was a small office in the department, and the newcomers were not around, they gossiped about the new staff. This small office was an important social space for interaction among the employees, which contributed to group cohesion. Although they all agreed that they required more personnel, many established staff members were angry that the management had not asked them for their opinion on this project and about their willingness to work with the newcomers. Birmasa, an established worker from Poland, thought that the newcomers required extensive support and guidance. These concerns were shared by others, but they indicated that there was no extra time for these duties and nobody was willing to sacrifice their free time. For the established staff, gossip was a way to protect and affirm their norms, beliefs, and professional relationships. This strengthened their internal cohesion but adversely affected their relationship with the newcomers. This phenomenon is consistent with Elias and Scotson's (1965) observations about the functions of gossip as a source of power that strengthens the ties within an already-cohesive established community, but also serves as a means to reinforce the position of newcomers as 'outsiders.'

A number of the newcomers were equally resentful regarding the manner in which they were treated by their established colleagues. Fadhil, an Iraqi newcomer, realized that the local staff was gossiping about them; this angered him. He recalled an instance when he was working with Sebastian, a German colleague, and asked for permission to leave work earlier. Sebastian granted him leave. But the next day, Fadhil received a warning from the management. Other (highly skilled) newcomers felt unappreciated and disrespected by the management, who acted as though their previous work experiences were irrelevant. Rash, for instance, was trained as an accountant in Syria and could not find a job related to his educational background in Germany. He was under the impression that, unlike German citizens, he lacked equal opportunities to secure desirable employment. In Germany, everything revolved around diplomas, but he had lost all his documents because of the Syrian war. He felt compelled to take on 'left over' jobs—a feeling that was confirmed when he saw a large number of migrants working in nursing homes.

A few of the newcomers were very suspicious of the management. Neram, an Afghan national, was uncertain about whether the management would stand behind him. He feared that should something unpleasant happen, he would be blamed. Asmal, a newcomer who had worked as a chef in several restaurants in Iraq and Lebanon, wondered why the management denied him a position in the kitchen. Similar questions were raised by Asmat; trained as a nurse in Afghanistan, he would rather work in a hospital than in a nursing home. These newcomers were pained by the idea that their wishes did not matter to the management.

In addition, many of the newcomers were insecure regarding their residency status and afraid that they might be sent back to their country of origin. For some, this was the main reason for participating in the AFP project. For example, Sema, a 40-year-old worker from Azerbaijan, participated in the project because her application for a residence permit had not yet been approved. She, her husband, and her 10-year-old son lived in constant fear of deportation. Sema hoped that by being hired by the nursing home, she would receive a residence permit.

However, the feeling of being ignored by the management was not present only among the newcomers. Bitter indignation and frustration also showed in the remarks of some of the younger established professionals. For example, Anna, a second-year student-nurse born in Germany, believed it to be unfair that she was compelled to share her supervisory time with the AFP students. Because of the intensive mentoring needed by the newcomers, she felt that no time was left for her guidance.

The established staff felt they were disrespected by both the management and society in general. One of them vented her feelings in the following manner: 'We are not appreciated at all! Not by society, not by residents, and not by their relatives. We receive a lot of complaints. They expect us to be gods who can do everything. But we are only human.'

German-Congolese Belo positioned himself as an 'outsider': 'I've been in Germany for many years. Germany effectively provides jobs, but is not good for me, you know. Even though I am married to a German, I will never be one of them.' He stated that in the end, his fate did not matter because he would never belong to German society. Yet, the next moment he identified himself with his established German colleagues: 'I work hard just like the Germans, you know. But they [newcomers], they are not good workers.' Here, Belo was shifting boundaries, now he belongs to 'us' and the newcomers are the 'outsiders.' Some of the established workers were migrants themselves, but regarded the newcomers as competitors. For example, because Zenobia, a Polish worker, had no qualification for her current position, she was concerned over losing her job due to increased competition from the newcomers. Nonetheless, she was confident that her attitude and behavior were much better than those of the new arrivals. From her perspective, she made no demands, had not refused any task, and worked as diligently as her German colleagues. A large number of the Eastern European workers considered themselves as belonging to 'us,' as better integrated people than 'them,' the new arrivals. The established health care workers, the native Germans, the migrant workers, and the new arrivals were drawing multiple, interrelated symbolic boundaries that went beyond binary oppositions along for instance ethnic or national lines.

5 Discussion

Our study revealed the crucial role of organizational constraints in the social dynamic between established staff and outsiders. In the nursing home, a lack of staff, a demanding workload, and lack of resources considerably exacerbated the tensions between the groups. In addition to this, different norms and ideas with regard to the meaning of good care, language difficulties, and gossip regarding the presumed lack of interest and motivation among the newcomers contributed to their stigmatization and sense of alienation.

In line with Elias and Scotson's (1965, 2002) theory, established staff members drew boundaries to positively distinguish themselves from the other group. They positioned themselves as 'us, the hard workers,' who had superior norms and 'saw' the work that was needed to be done. In contrast, the newcomers were 'them, the outsiders,' who lacked motivation, had language and skill deficiencies, and failed to 'see' the work. The established migrant workers deemed themselves 'better integrated people' than the new arrivals. In addition, given their duration of stay in the field, with a few having acquired their caring qualifications in the German system, they had familiarized themselves with the habitus and the rules and routines in the ward and developed an innate comprehension of the intricacies of their work (Bourdieu, 1990).

However, seen through a Foucauldian lens, the gossiping among the established staff can also be interpreted as a form of resistance, or a 'weapon of the weak' (Ginzberg, 2017; Scott, 2008). In Foucault's conception, power works by normalizing discourses and this is embedded in every institution and in every relation as part of everyday practice, thereby subjecting all workers, whether established or newcomer, to 'disciplinary power' (Foucault, as cited in Hutchinson et al., 2006). This means that in different contexts people hold different positions, but there is not one particular group that dominates over or possesses more power than another group. In other words, power is omnipresent, because it is produced from moment to moment in every relationship (Foucault, 1978). The findings of the current study are more in line with this conception of power as a relational and dynamic force.

Although the established workers mostly dominated and had more power because they coordinated the tasks in the wards, occasionally, the newcomers also showed signs of resistance, for example when they refused to perform particular tasks or when they followed their own ideas with regard to what constitutes good care by investing more time in caring for a resident than was routinely done or expected. Thus, their power, like gossip and resistance, is more in line with what Hutchinson and Jackson (2015) described as an 'imaginary sense of power.' What they all had in common was that they felt they were not appreciated and respected (not by their management and, more generally, by society).

In scrutinizing the interaction between established workers and newcomers in health care institutions for the elderly, the theory of Elias and Scotson has two further drawbacks. First, the established–outsider figuration can easily lead to the assumption that the boundaries between these groups are fixed, static, and solid (Wimmer, 2013). However, such a binary, oppositional categorization of 'us' versus 'them,' obscures the numerous possible points of affiliation that are often forged across these lines. Symbolic boundaries are always socially constructed (Wimmer, 2013). In other words, distinctions that separate people into different groups and categories easily gloss over the dynamic and diverse process of boundary making, wherein who belongs to whom is often unclear because categories may be fuzzy, fluid, and malleable. The participants in this study used a variety of markers, to which they attached different meanings related to social interaction. A few of the employees in the nursing home made distinctions across different groups and occasionally they blurred the boundary between 'us' (established care workers) and 'them' (newcomers). For example, the established Congolese worker distinguished himself as a 'hard worker' who belongs to 'us' and categorized the newcomers as 'them,' who are 'not good workers.' When addressing the diversity in general in Germany, he re-positioned himself as 'outsider' as he believed that he would never belong to the established group in society.

Second, Elias and Scotson's study was limited to the internal dynamics within a suburban community, where the role of the wider context and relations with the larger society were disregarded (Hogenstijn et al., 2008; Petintseva, 2015). In applying their framework to cast light on the social dynamics between established health care workers and newcomers in a home for the elderly, the risk is that the effects of the institutional and national context are ignored. Moreover, we found that the dynamics between the groups were significantly influenced by institutional factors, such as work pressure, institutional standards of care, routinized ways of working, and organizational constraints.

6 Conclusion

This study highlighted the social dynamics between established health care employees and newcomers in a nursing home for dementia residents in Germany and how concepts of good care affected the interaction between these two groups. This investigation also identified the role of various organizational constraints, such as shortages in staff, a demanding workload, and a lack of resources. All the participants experienced tensions, irritations, and frustrations in daily practice. The established employees maintained control, self-respect, and their positions in the caregiving team by maintaining a distance from the newcomers. Further, a few of the established health care employees were insecure about the potential economic, cultural, demographic, and political effects of the arrival of a new group of workers, whereas others who had a migrant background themselves and/or lacked formal qualifications feared that the newly arrived 'strangers' would surpass them. On the part of the newcomers, there were those who felt frustrated owing to the lack of opportunities for which they aspired. These processes supported and reinforced group divisions in the nursing home, thereby giving rise to collective images of 'us versus them.' This, in turn, negatively influenced the interactions between the established staff and newcomers, despite the cultural awareness and sensitivity training provided to established staff.

This study has several empirical limitations, which could be addressed in further research. First, additional research is needed into the micro-level of personal interactions between newcomers and residents, newcomers' sense-making processes, and the effects of these two aspects on elderly care (intergroup relations). Second, it would be interesting to investigate, at the meso-level, whether institutional initiatives like the AFP project can develop into a sustainable and long-lasting commitment of nursing homes (organizational and institutional schemes). Third, this would require focusing on the macro-level as well, because the position of asylum seekers and refugees is influenced by the tenor of politics and policy and public debate at both the national and international levels.

Acknowledgments I would like to thank all participants of the nursing home in Germany. For ongoing conversations and for feedback on this article, I am also very grateful to Dr. Baukje Prins and Prof. Dr. Ria Reis. I would like to thank the editors for their constructive feedback and their invitation for publication.

Funding This study was funded by the Dutch Organization for Scientific Research (NWO) (023.008.044/1105).

Notes

1. A society is deemed super-aged when the population of individuals 65 years or older exceeds 21% of the total population (FSO, 2016).
2. The term 'refugee crisis' refers to migration mobilization development. Some scholars claimed that this mobilization is better depicted as 'a crisis of European border regime,' 'a crisis of refugee protection,' or a 'European political crisis' (Geddes, 2017; Scherr & Yüksel, 2016; Schwiertz & Ratfisch, 2016).
3. The definition of workforce integration is adapted from Ramji and Etowa's (2018, p. 15) description: The integration in the professional environment resulting from mutual adjustments between immigrant health care workers and established care personnel.
4. The training of an *Altenpfleger/in* (geriatric nurse) lasts three years after vocational training; the training of a *Pflegerassistent/*in (nurse assistant) or *Altenpflegerhelfer/in* (care assistant) lasts one year; and the training of an *Alltagsbegleiter für Menschen mit Demenz* (nurse aide) lasts 12 weeks. See also: https://www.eurodiaconia.org/wordpress/wp-content/uploads/2016/08/The-education-training-and-qualifications-of-nursing-and-care-assistants-across-Europe-Final.pdf.
5. The content of the intercultural competence program for the established staff was identity and factors influence human identities, biography, diversity in health care, explanatory models (illness, disease, sickness), medical

pluralism, caleidoscopia game, intercultural communication, TOPOI model, prejudices and stereotypes, and geography and citizens in town.

6. Scabies is highly contagious and spread mainly through direct skin-to-skin contact. Symptoms tend to present after 4–6 weeks upon exposure or within a week after repeated infestation. Scabies appear as a skin rash and pruritus, which typically worsens at night.
7. Germany requires newcomers who apply for citizenship or permanent residency to prove that they have achieved a B1 level of German language proficiency, which stands for 'intermediate' in the Common European Framework of Reference for Languages. Many employers in the skilled labor market require at least level B2 (upper-intermediate) German language proficiency.

References

Bonin, H., Braeseke, G., & Ganserer, A. (2015). Internationale Fachkräfterrekrutierung in der deutschen Pflegebranche. Retrieved January 14, 2020, from https://www.bertelsmannstiftung.de/fileadmin/files/Projekte/28_Einwanderung_und_Vielfalt/Studie_IB_Internationale_Fachkraefterekrutierung_in_der_deutschen_Pflegebranche_2015.pdf

Bourdieu, P. (1990). *The logic of practice*. Stanford University Press.

Bundesministerium für Gesundheit. (2018). Beschäftigte in der Pflege. Retrieved June 8, 2020, from https://www.bundesgesundheitsministerium.de/themen/pflege/pflegekraefte/beschaeftigte.html

ECRE. (2020). Statistics Germany. Retrieved January 7, 2020, from https://www.asylumineurope.org/reports/country/germany/statistics

Elias, N. (1994). Introduction – A theoretical essay on established and outsider relations. In N. Elias & J. L. Scotson (Eds.), *The established and the outsiders – A sociological enquiry into community problems* (pp. xv–lii). Sage.

Elias, N., & Scotson, J. L. (1965). *The established and the outsiders – a sociological enquiry into community problems*. Sage.

Elias, N., & Scotson, J. L. (2002). Weitere Facetten der Etablierten-Aussenseiter-Beziehung. Das Maycomb-Modell. In N. Elias & J. L. Scotson (Eds.), *Etablierte und Aussenseiter* (pp. 285–308). Suhrkamp.

Federal Statistical Office. (2015). 20% der zugewanderten Pflegekräfte stammen aus Polen. Retrieved February 4, 2020, from https://www.presseportal.de/pm/32102/3076454

Federal Statistical Office. (2016). Older People in Germany and the EU. Retrieved February 4, 2020, from https://www.bmfsfj.de/blob/113952/83dbe067b083c7e8475309a88da89721/aeltere-menschen-in-deutschland-und-in-der-eu-englisch-data.pdf

Foucault, M. (1978). *The history of sexuality, volume 1: An introduction.* Allan Lane.

Geddes, A. (2017). How to reassure the 'anxious middle'? Reflections on the drivers of migration and of migration politics. In F. Pastore (Ed.), *Beyond the migration and asylum crisis* (pp. 103–107). Aspen Institute.

Ginzberg, E. (2017). Weapons of the weak or the culture of everyday resistance to power relations. *GSTF International Journal of Law & Social Sciences, 3*(2), 1–14.

GIZ. (2015). Gewinnung von Pflegekräften aus dem Ausland – unser nachhaltiger Ansatz. Retrieved June 8, 2020, from https://www.giz.de/en/downloads/TripleWin_Factsheet_2016.pdf

Hanewinkel, V. (2015). Ein Interview mit Dr. Helma Lutz. Auslandische Plegekrafte in deutschen Haushalten. Retrieved January 25, 2020, from https://www.bpb.de/gesellschaft/migration/kurzdossiers/211011/interview-mit-helma-lutz

Ham, A. (2020). First generation immigrant and native nurses enacting good care in a nursing home. *Journal of Nursing Ethics*. https://doi.org/10.1177/0969733020921487.

Ham, A. (2019). Social processes affecting the workforce integration of first-generation immigrant health care professionals in aging citizens in the Netherlands. *Journal of Transcultural Nursing*. https://doi.org/10.1177/1043659619875196.

Hogenstijn, M., Van Middelkoop, D., & Terlouw, K. (2008). The established, the outsiders and scale strategies: Studying local power conflicts. *The Sociological Review, 56*(1), 144–161.

Horn, V., Schweppe, C., Böcker, A., & Bruquetas Callejo, M. M. (2019). Live-in migrant care worker arrangements in Germany and the Netherlands. Motivations and justifications in family decision-making. *International Journal of Ageing and Later Life*. https://doi.org/10.3384/ijal.1652-8670.18410.

Hutchinson, M., & Jackson, D. (2015). The construction and legitimation of workplace bullying in the public sector: Insight into power dynamics and organizational failures in health and social care. *Nursing Inquiry, 22*(1), 13–26.

Hutchinson, M., Vickers, M., Jackson, D., & Wilkes, L. (2006). Workplace bullying in nursing: Towards a more critical organizational perspective. *Nursing Inquiry, 13*(2), 118–126.

Kingma, M. (2018). *Nurses on the move: Migration and the global health care economy*. Cornell University Press.
Kniejska, P. (2015). All-inclusive-Pflege aus Polen in der Schattenzone. *WISO direkt.* Retrieved January 25, 2020, from https://library.fes.de/pdf-files/wiso/11341.pdf
Petintseva, O. (2015). Approaching new migration through Elias 'established' and 'outsiders' lens. *Human Figurations, 4*(3), 1–15.
Rada, A. (2016). Migration of health-care workers from the new EU Member States to Germany. Major trends, drivers and future perspective. Retrieved June 12, 2020, from https://www.beobachtungsstelle-gesellschaftspolitik.de/f/915ef8f93d.pdf
Ramji, Z., & Etowa, J. (2018). Workplace integration: Key considerations for internationally educated nurses and employers. *Administrative Sciences, 8*(1), 2–18.
Scherr, A., & Yüksel, G. (2016). Flucht, Sozialstaat und soziale Arbeit. *Neue Praxis* (Sonderheft 13).
Scott, J. C. (2008). *Weapons of the weak: Everyday forms of peasant resistance*. Yale University Press.
Schwiertz, H., & Ratfisch, P. (2016). Rassismus und anti-migrantische Bewegungen im deutscheuropäischen Migrationsregime. In S. Hess et al. (Eds.), *Der lange Sommer der Migration: Grenzregime III* (pp. 151–162). Assoziation A.
Ulusoy, N., & Schablon, A. (2020). Discrimination in in-patient geriatric care: A qualitative study on the experiences of employees with a Turkish migration background. *International Journal of Environmental Research and Public Health, 17*(7), 2205.
World Health Organization. (2020). Dementia. Retrieved February 10, 2020, from https://www.who.int/health-topics/dementia#tab=tab_1
Wimmer, A. (2013). *Ethnic boundary making: Institutions, power, networks*. Oxford University Press.
Xiao, L. D., Willis, E., & Jeffers, L. (2014). Factors affecting the integration of immigrant nurses into the nursing workforce: A double hermeneutic study. *International Journal of Nursing Studies, 51*(4), 640–653.
Zhou, Y., Windsor, C., Theobald, K., & Coyer, F. (2011). The concept of difference and the experience of China-educated nurses working in Australia: A symbolic interactionist exploration. *International Journal of Nursing Studies, 48*(11), 1420–1428.

Part III

Risks and Social Protection

10

Everyday Vulnerability: Work and Health Experiences of Live-in Migrant Care Workers in Taiwan

Li-Fang Liang

1 Introduction

The force of globalization not only contributes to flows of capital and goods but also facilitates the circulation of people. Worldwide, approximately 258 million migrants live in countries in which they do not hold citizenship. Migrant workers, people working or seeking work in their present country of residence, comprise 61.5 percent of them, some 164 million.

Among these, 41.6 percent, 68 million, are women (Popova & Özel, 2018). This proportion is likely to increase as migration is increasingly feminizing, as reflected in the gender of migrant workers and the gendering of the work they perform. Countries in East and Southeast Asia host a huge number of migrant domestic workers and care workers, most of them women (Miller & Castles, 2009), and Taiwan has hosted such workers since 1992.

L.-F. Liang (✉)
Department of Sociology, National Dong Hwa University,
Taipei, Taiwan

V. Horn et al. (eds.), *The Global Old Age Care Industry*,
https://doi.org/10.1007/978-981-16-2237-3_10

Domestic work and care work that migrants perform in more developed countries are perceived as dirty, difficult, and dangerous, reflecting institutionally and socially constructed constraints in both sending and receiving countries (Huang & Yeoh, 2003). In the Taiwanese context, most migrant care workers work in private households, although the country also employs migrants in medical institutions, such as nursing homes. These workers provide personal care, sometimes including direct health care, to older people and people with disabilities but are also asked to do household chores (e.g., cooking and cleaning) and to provide childcare.

During recent decades, research on migrants and health has moved beyond the individual behavioral approach to examine how social structure and everyday violence shape particular experiences of different individuals or social groups. For example, Holmes (2007, 2013) and Benson (2008) have researched migrant farm workers in the United States, demonstrating that unequal social structures injure workers physically and mentally. They illustrate that the naturalization of structural violence causes intimate suffering for migrant workers through everyday practices of employment, law enforcement, and liberalized production regimes, which market forces drive. Walter and his colleagues' ethnographic work (Walter et al., 2004) on Latino day laborers provides data on social suffering and embodies the linkages between structural forces and intimate suffering at the individual level. However, most work focuses on migrant workers in agriculture (e.g., Arcury & Quandt, 2007; Mayell & McLaughlin, 2016) and pays little attention to other sectors, including care work.

Household labor is invisible (Anderson, 2000; Ehrenreich, 2002) due to its unregulated and gendered characteristics. However, a growing body of literature on live-in migrant workers, including care workers and domestic workers, draws attention to their working conditions and violations of their employment rights (Datta et al., 2006; Khan, 2009; Walsh & O'Shea, 2010; Stevens et al., 2012; Atanackovic & Bourgeault, 2014; da Conceição Figueiredo et al., 2018). These studies contribute to our understanding of how the institutional context, including migration policy, migratory status, and work environment and conditions, affects migrant live-in workers' daily experiences. However, health issues have remained under-investigated.

The research on care work demonstrates that care work is not only physically demanding but also emotionally taxing (Brotheridge & Grandey, 2002). In addition to occupational hazards, studies of care workers and family caregivers reveal the multiple factors that make caregiving stressful and dangerous to mental health (Silva-Smith, 2007; Bialon & Coke, 2012; Paradise et al., 2015). Compared to care workers in general, live-in migrant care workers have a more vulnerable social position because of the intersectionality of gender, race/ethnicity, class, migratory status, and the live-in condition (Yeoh & Huang, 2010; Lutz, 2011; Pratt, 2012). However, few studies address the health and well-being of live-in migrant care workers (see Ahonen et al., 2010 for an exception). These studies reflect the occupational medicine approach, and therefore they focus on work conditions and environment. For example, Ahonen and her colleagues' research (2010) documents occupational hazards faced by migrant care workers and their effects on mental health. Although this approach shifts the determinate factors of health from individual to work environment, it fails to address the broader institutional context beyond work settings.

Extending beyond the occupational medicine approach, the current research not only tries to understand the relationship between employment and health experiences, but also to include the broader structural and institutional context in which care workers labor. In this chapter, I adopt the concept of structural violence (Galtung, 1969) to examine the structural constraints and factors that shape the health and working experiences of migrant live-in care workers. Taiwan's guest worker scheme, which was created in 1989, grants migrant workers temporary working visas and thereby makes them more vulnerable than permanent visa holders would be. In examining the consequences, I pay particular attention to work-related stress and mental well-being.

The concept of structural violence is rooted in the liberation theology of Latin America. Galtung (1975) defined structural violence as "the indirect violence built into repressive social orders creating enormous differences between potential and actual human self-realization" (p. 173). Taking up the idea, medical anthropologist Paul Farmer extends the concept of structural violence through developing an ethnography of socially structured inequality, which examines the causal linkage between macro-dynamics and individuals' lived experiences (Farmer, 2004). Many scholars in social science have utilized the concept to analyze the influences of socially structured inequality on individuals' intimate experiences. The concept has been applied to investigate medical and health issues. The collective experiences of social suffering embody the systematic constitution of inequalities that political–economic processes, social structures, and cultural ideologies shape (Benson, 2008).

The remainder of this chapter proceeds as follows. First, I introduce migrant care labor policy in Taiwan to situate the group of live-in migrant care workers in the broader institutional context. Second, I describe this study's design and methods. The next section examines the liminal status of live-in migrant care workers and how it shapes their working conditions, employment relations, and health experiences to illuminate the structural violence embedded in their daily lives. The fourth section examines how employers' and live-in migrant care workers' focus on their respective, immediate economic interests prevents them from prioritizing migrants' health in this regard. To conclude this chapter, I argue that governments in labor-receiving countries have an important role to play in relation to protecting the rights of live-in care migrant workers, including their right to health.

2 Live-in Migrant Care Workers in Taiwan

Taiwan officially became an aging society in 1993, when its population of adults older than 65 reached 7 percent of the total population. By the end of 2019, 15.28 percent of the population was senior (Ministry of the Interior, 2020), and official counts suggest that it will increase to 20 percent by 2026, reflecting declines in fertility rates and longer life spans (National Development Council, 2018). Compared with Western societies, Taiwan's "graying" rate has been extremely rapid. For example, France's older adult population increased from 7 percent to 14 percent over the course of 115 years, more than quadruple the time it has taken in Taiwan. The resulting rise in demand for long-term care in Taiwan is dramatic.

Since 1992, the Taiwanese government has allowed the immigration of domestic workers and care workers as part of a short-term contract labor force. These workers care for older adults, people with disabilities, the sick, and young children. Families hiring live-in migrant care workers have to receive approval from the Ministry of Labor based on a subjective and problematic needs assessment of the cared-for (Liang, 2015). Earlier research on live-in migrant workers in Taiwan showed that employing a live-in worker is a privilege restricted to middle and upper-middle class families (Lan, 2006). However, the socioeconomic background of employers has become more diverse over time. The monthly salary of live-in migrant care workers is about US $667, not including overtime payment for working on Sundays. For many families with care needs, hiring a live-in migrant care worker is more affordable, flexible, and culturally acceptable than alternative care arrangements such as institutional care (Liang, 2017).

The number of live-in migrant care workers in Taiwan has increased rapidly. The Ministry of Labor (2020a) provides ample statistics on migrant workers in the country. In 1992, there were 306; by the end of 2019, there were 244,379 (there are also 15,281 institutional migrant care workers and 1797 live-in domestic migrant workers not specifically charged with care provision). The country imports migrant workers for non-domestic purposes as well, but live-in care workers comprise more than one-third of all the migrant workers in Taiwan. The majority, 197,385, come from Indonesia. An additional 29,046 are Filipina, 17,536 are Vietnamese, and 385 are Thai. More than 99 percent of the migrant workers in care and domestic worker positions are women (Ministry of Labor, 2020a). All of these sending countries have signed a bilateral agreement with the Taiwanese government. Thus, it is not immediately clear why Indonesian women dominate so heavily. Whether as a cause or a consequence, private employment agencies that act as brokers connecting migrant care workers with families have mobilized a gendered-ethnicized discourse claiming that Indonesian women are more submissive and obedient than others, thereby portraying them as ideal for the work live-in care workers perform (Liang, 2011).

Although the government opened the channel of direct hiring for live-in care workers working for the same contracted employers in 2008 (for industrial workers in 2009), most employers still rely on private employment agencies to match, hire, and manage workers (Liang, 2011). The private employment agencies are among the important actors facilitating and conditioning labor migration. They make migrants' journeys easier to arrange but more expensive for them (Xiang & Lindquist, 2014). Most migrant workers pay a huge service fee to the private employment agencies in their home countries. The Ministry of Labor has issued Standards for Fee-Charging Items and Amounts of the Private Employment Services Institution,[1] according to which migrant workers must pay a monthly service fee to Taiwanese employment agencies throughout their work period in Taiwan. This fee declines over the three years for which they hold their visas, from US $60 to $55 to $50.[2] The Regulations for Permission and Supervision of Private Employment Services Institution[3] requires employment agencies to provide services to migrant workers, including ensuring they have a place to live; providing consulting, counseling, and translation services; and arranging their entry, departure, and periodic health check-ups and reporting the results to the government's health authority. However, in practice employment agencies primarily function to help employers manage and discipline migrant workers (Cheng, 2003; Liang, 2011).

Most workers cannot work in Taiwan more than 12 years; live-in care workers can work for 14 years. After that, they are not permitted to enter Taiwan on working visas again. The government emphasizes the temporary status of migrant workers, claiming they should be supplemental to local labor, rather than substituting for it. Migrant workers do not have the same employment rights as native Taiwanese or migrants who come to the country for white-collar jobs. Their freedom of mobility is limited. Not only must they hold a government-issued permit but they cannot work for an employer who does not also hold a permit. Additionally, they are not allowed to change between occupational sectors; for example, a live-in care worker cannot switch to institutional care or industry. Live-in care workers do not have any choice in their accommodations—they have to live with those they care for. Most are employed by the adult children of the care recipients.

The Taiwanese government argues that, because of its limited territory, Taiwan cannot have a lenient immigration policy. Migrant workers are not allowed to bring their families with them or to get married (to either locals or migrants) while they are working in Taiwan. For more than a decade, the state administered pregnancy tests to female migrant workers and denied those who tested positive entry until the practice was outlawed in 2002. However, in reality the recruiting agencies and employers have power over workers' reproductive rights. While national employment law prevents employers from dismissing workers because they are pregnant, employers can claim that a pregnancy has made a worker unable to fulfill her job duties or force workers to resign when they are pregnant.

The Taiwanese government does not regulate the working conditions of live-in migrant care workers. Live-in workers do not benefit from the Labor Standards Act's minimum wage, overtime pay, or number of days off requirements. Without legal protection, these migrant workers are at a high risk of abuse and sexual harassment, due to the unequal power relations exercised through the binary of domination and subjugation, their live-in arrangement, and the intersectionality of gender, class, ethnicity, and migratory status.

3 Research Design and Methods

The research aims to explore how the structural constraints that are embedded in the daily lives of Indonesian live-in migrant care workers affect their work and health experiences. I conducted a pilot study consisting of two five-person focus groups with live-in migrant care workers to ensure that the interview questions we asked in the main study would be comprehensible to research participants. I revised some questions based on the results. The dimensions of the interview questions included personal information and previous work history, the motivation of migration and process, work experiences and conditions, health-related issues, and social and family support. The data collected through the focus groups is included in the analysis.

The main study consisted of 25 individual interviews with live-in migrant care workers. About half of the interviews were conducted when the workers accompanied their care recipients to public parks and community centers. The rest were completed either when the workers had a day off, typically in public places, or in the homes where they worked or at their employers' houses. No employers were present at the time of the interviews, reflecting my concern that power relations would prevent workers from telling me about their actual experiences. Since the elderly they care for typically have serious disabilities, illnesses, or dementia, I did not feel their presence would have the same chilling effect. In addition to the workers' interviews, I also draw, here, on interviews I conducted for an earlier study with 35 employers (10 care recipients and 25 family members, including the sons, daughters, daughters-in-law, and spouses of care recipients).

I recruited the research participants for both phases of the study through our own social networks, with the help of the Catholic-affiliated Organization for Migrants and snowball sampling. The organization holds Mandarin classes every Sunday and provides religious services for migrant workers of all religious affiliations. After each interview, I asked the interviewee for names of additional potential participants. An Indonesian translator assisted with participant recruitment and data gathering. She works professionally as a translator for governmental and business sectors. Some participants have worked in Taiwan for a long time and are fluent in Mandarin, but the translator and I worked together on all interviews, explaining the study's goal to research participants before each interview began and obtaining informed consent for participation as well as, in most cases, permission to record their interview. For those interviewees who were not comfortable with the recorder, we took notes during interviews and later developed them into detailed transcriptions.

Most of the participants were from rural Java; one was from metropolitan Jakarta and two were from Sumatra. Their ages ranged from 26 to 46 years old. Only two had never been married and had no children, while the rest were either married or divorced with one to three children. All had formal schooling; the majority had finished junior high school at least and some had completed high school. Before coming to Taiwan, most of them had gained experiences of working overseas as live-in domestic workers or factory workers in Singapore, Malaysia, and the Middle East. More than two-thirds of them had worked in Taiwan for more than six years.

4 Liminality of Live-in Migrant Care Workers

Taiwanese migrant labor policy offers no path to citizenship. If migrants have children in Taiwan in spite of prohibitions on marrying and infringement on their reproductive rights, their children will not be citizens, as Taiwan lacks birthright citizenship. Migrant workers' legal status is bound to their employment with specific employers; they can only take jobs with employers who have applied for the right to hire them. They cannot even change from one employer to another employer entitled to hire a migrant worker without the approval of the Taiwanese government. This regulation contributes to migrant workers' dependency on their employers and aggravates their unequal power relations with them. In this section, I develop the analysis through the lens of structural violence, which demonstrates how structural force relates in an embodied manner to individuals' daily experiences. The physical illnesses and problems with mental well-being live-in migrant workers experience are not personal issues. Instead, the migrant labor regime, employment experiences, and power relations exercised in the intimate sphere through providing and receiving care work shape their sufferings.

4.1 Long Working Hours

Absent the protections of Taiwan's Labor Standards Law, private contracts dictate the working conditions of these live-in migrant care workers without government regulation. They can be required to work 24 hours a day. While a governmental survey of employers (Ministry of Labor, 2020b) puts the average working day of live-in care workers at 10.4 hours and states that 34.4 percent of them did not have any days off, my interviews suggest that these are underestimates. For example, Keyla, an Indonesian live-in care worker, was 31 years old. She had worked as a live-in care worker in Taiwan for almost 11 years, starting as soon as she graduated from high school. She described the conditions at a job she had for years taking care of an elderly couple thus:

> I had no days off. I provided round-the-clock care to [my female care recipient]. I could not leave her alone. We slept in the same bed. I could not have enough rest until she passed away. … We were very close. When she was hospitalized, I accompanied her in the hospital. When she was sick, I was sick too. [Her family] wanted to hire another care worker who could stay in the hospital with her [so Keyla could stay with her male care recipient], but she rejected that. She begged me [to go with her].

It is not uncommon for live-in migrant care workers to care for both spouses when working in a household with a married couple. Siti was in charge of the daily life of an elderly couple in their early 80s. She signed a contract before coming to Taiwan indicating she would work 365 days a year. She never went out alone. She described relying on her employer, his wife, or her friend, a live-in migrant care worker next door, to purchase necessities for her that were not available at the grocery store nearby.

The nature of care work creates huge demands for physical and emotional labor. Most live-in migrant care workers provide care to the elderly who need assistance with activities of daily living. Many of them work long hours to provide hands-on care as well as emotional support. The workers' daily lives are organized to meet and satisfy the needs of the cared-for. The boundary between work hours and private time is blurred. Annie, who had worked in Taiwan for almost six years, said: "I did not have explicit working hours. We did not have the regulation of that. I had to provide help whenever [my charges or their family] needed it. If they did not go to sleep, I did not dare to sleep." Annie was 25 when she first came to Taiwan from Indonesia. The woman she took care of was frequently restless at night.

Wati had worked in Taiwan almost 12 years when we met. She described an experience similar to Annie's. She said that she had almost never had enough rest since she worked in Taiwan. She lived alone with an elderly couple and their adult children seldom visited them, so she got no relief. Like the other live-in care workers I spoke with, she used "A-Ma" ("Grandmother" in the Taiwanese dialect) and "A-Kong" ("Grandfather") to refer to her care recipients, which are Taiwanese terms of respect for older people not necessarily connoting kinship. As she said:

> I slept with the A-Ma and A-Kong in the same room. Their daily schedules were very different. The A-Ma usually woke up at two or three a.m. while the A-Kong was sleeping soundly. If I did not satisfy her request immediately, she would make noises by knocking on the wall. I was used to sleeping a few hours per day. I was used to it.

The A-Ma had passed away, but Wati still had little relief. Most live-in care workers described sleeping with their care recipients in the same room. Sharing a bed was common. Many interviewees mentioned that they rarely slept through the night without interruption. Some of them needed to assist the cared-for in either changing their diapers or going to the restroom at night and they had to respond to other requests as well. Others were unable to sleep soundly because of the noises care recipients made during the night, and those who cared for the elderly with dementia were particularly likely to be disturbed at night.

The government stipulates that migrant care workers hired by individual families have to live with their care recipients. This live-in arrangement makes it easier for employers to require long hours. It also makes workers vulnerable to physical and other kinds of abuse (Cheng, 2003; Liang, 2014). Even without abuse, as Annie said: "We were not part of the family, but we had to live together. … If I did not think of the salaries, I would run away. I often felt uncomfortable when I lived with [my employer and his family]. I always had to obey them."

Live-in migrant care workers' work hours and lack of days off have serious impact on their health and well-being. The live-in care workers we interviewed often used the adjective "exhausted" to describe themselves. Reni had worked in Taiwan almost nine years without regular days off. She looked exhausted every time when I met her. During our last interview, she expressed her longing to go home. I asked her why, and she just emphasized: "I am tired."

4.2 Heavy Care Responsibilities

In addition to long work hours, many interviewees described stresses due to their care responsibilities. While most had worked outside of their home country before coming to Taiwan, most had no experience caring for the fragile elderly. Recruiting agencies for old-age care workers in Indonesia provide minimal training prior to placement (Liang, 2011). Amanna described her life when she cared for an elderly woman for six months:

> I was very nervous all the time. I felt that I was always in the status of being ready to work. I never dared to sit down or relax. I always stood up for my employer to call me. The A-Ma I took care of was quite old. She liked to walk back and forth in the house. I followed her closely and was careful of her. It was very dangerous if she fell down.

Amanna had worked as a babysitter in her home country. She emphasized that caring for the elderly was extremely different from caring for a baby. She did not know what old-age care would require when she came and she found the learning process stressful.

Because of the limited training provided in the training center, like Amanna, Mae learned how to take care of the fragile elderly on the job. Even though she had worked in Taiwan more than ten years, she was still developing new care skills as her care recipient's health condition changed. When the woman she cared for was hospitalized because of heart failure, Mae had to learn how to extract sputum for her, because she would have to operate the apparatus by herself at home after the care recipient was discharged. I asked Mae if she was worried about or was afraid of extracting sputum. She smiled bitterly, saying: "In the beginning, I was worried about hurting the A-Ma. But I don't have any other choice. I am the only person taking care of A-Ma daily. I have to [extract sputum for her]."

It was stressful for Mae to provide the round-the-clock care with limited resources and support. Although her employer always expressed gratitude to her, this did not alleviate Mae's workload or stress.

Wati recalled how she learned care skills from the community nurse who visited her cared-for every two weeks. Even after caring for A-Kong a long time, she still felt stressed due to the uncertainty and degeneration of his health condition. As she explained:

> When I started taking care of the A-Kong, his health was not as bad as how he is now. He had suffered a stroke. But he was able to walk slowly with a walker. In the past years, I saw that his health was declining. Now, he is bedridden and relies on me to turn over, eat, take a bath, and so on. I cannot sleep well at night. I am afraid that his health condition will suddenly get worse.

Many workers we interviewed shared similar experiences and feelings, and those who lived with the cared-for alone showed particular stress. Witnessing decline of their care recipients' physical and mental health took a toll. As the next section describes, many felt strong bonds with their care recipients and felt responsible for their well-being even as it inevitably deteriorated. Emotional ties and their mutual marginal position in the broader society facilitated closeness (Liang, 2018). Workers' uncertainty about their future in Taiwan also played a role.

4.3 The Power Dynamics in Care Workers' Employment Relations

Taiwan's migrant labor policy places live-in migrant care workers in a liminal space where their employers manage and surveil their lives (Cheng, 2003; Liang, 2011, 2014). Many employers described care workers as family members, but this seemed to compound their control (Anderson, 2007; Frantz, 2008; Lin & Bélanger, 2012). Keyla said:

> I did not dare to tell [my employers, the son and daughter in-law of the cared-for] that I was tired. Sometimes I wanted to tell [them] that it was unreasonable for me to do all the work [they demanded of me]. ... They always said: 'you are part of our family.' But what did they mean by being a family member?

The kinship analogy adopted by some employers presents a paradox. On the one hand, employers recognize the workers' hard work and their devotion to the cared-for. On the other hand, sometimes the analogy justifies exploitation within the relationships and a demand that workers be loyal to their employers, care recipients, and other family members to the extent of fulfilling any task, regardless of the cost to themselves. Although many interviewees emphasized that they had benign employers, they understood the power dynamics between themselves and their employers and seldom disobeyed their employers' demands or failed to enact their preferences.

Annie was an exception. She joined the Association of Indonesian Migrant Workers in Taiwan (ATKI-Taiwan), in spite of her employer's disapproval. She emphasized that the Taiwanese government should regulate working hours for live-in migrant care workers. Her employer was not pleased with her participation, but did not forbid her or fire her. Yet in many ways, Annie remained docile. The expectation of submissiveness has deep roots (Liang, 2011). Annie had been trained at a recruiting center in Indonesia before applying for a job in Taiwan. She pointed out that the main goal of training courses was transforming every unique individual into a docile and obedient worker:

> When I was in the training center, I was not allowed to have a different opinion from the recruiting agency. No matter what I would like to do, I had to finish the work assigned. Otherwise, I was not qualified to work overseas. [In Taiwan] the situation was the same. I had to listen to my boss or quit the job.

Annie said she was frequently unhappy. She had little autonomy as a live-in migrant care worker. Like Annie, many interviewees expressed that the homes where they worked, as well as their employers' homes, had become places of subordination. They had to listen to and obey their employers.

Workers did describe their relationships with care recipients as family-like, however. Emotional ties and interdependency can complicate power dynamics in relationships (Ayalon, 2009; Porat & Iecovich, 2010; Timonen & Doyle, 2010; Näre, 2011; Liang, 2018; Teshuva et al., 2019). Mae had cared for one woman for years, and she described their relationship as very close. While, as I have mentioned, A-Ma is a term of respect that doesn't necessarily imply kinship, Mae used it consistently, which seemed to suggest genuine feeling. However, her concern contributed to her isolation. While for years she had been able to take the A-Ma out using a wheelchair and thereby meet up with other Indonesian care workers in the park, when we spoke she was almost completely confined because the A-Ma's health had worsened. As she said, "I seldom go out now. The farthest place I go is the Indonesian store nearby. It's only three minutes walking distance. Sometimes, I go there to meet up with my friends when the A-Ma is taking a nap. Usually I don't stay long. I worry about the A-Ma."

Mae pointed out that the A-Ma is close in age to her own parents, and this seemed to strengthen the emotional tie she felt. Weiwei also described identifying the woman she cared for with a member of her own family: "A-Ma's age is quite similar to the age of my own grandmother in Indonesia. When I take care of her, it always reminds me my own grandmother. I feel that I am caring for my own grandmother." Weiwei had cared for the woman, who had dementia, for four years.

For Weiwei, a sense of shared marginalization played a role as well: "She is too fragile to take care of herself. She has to rely on me. Her sons and daughters-in-law have their own work. They don't have time to care for her." She often said, "A-Ma and I are together," and "If I were not there, who will take care of my A-Ma?" However, she also mentioned that her employers, the eldest son in the family and his wife, showed little concern for her well-being. Weiwei is Muslim and she does not eat pork. But her employers often gave her a bun stuffed with pork meat as her breakfast.

Several care workers, including Mae and Weiwei, attached complex meanings to the care work they provided. They expressed affection and obligation as well as burden (Liang, 2018; Teshuva et al., 2019). Their sense of responsibility to care recipients went beyond a job. It was driven by their emotional ties with the cared-for and the relationships of interdependency between these two parties. The boundary of formal work was blurred. However, this did not protect them from the toll of full-time care work.

5 A Healthy Body as a Workable Body

Live-in migrant care workers usually choose to endure their harsh working conditions and the related illness because of their economic needs and hope for a promising future. In addition, unequal power relations between themselves and their employers guide their actions and behaviors. Within these contexts, they do not prioritize their health. Employers often regard live-in migrant care workers as usable labor rather than as individuals with physical, reproductive, and emotional needs (Lindio-McGovern, 2003). They fail to recognize that migrant live-in care workers are human beings situated in their specific social and historical contexts (Chang, 2016). In this section, I illuminate why and how the live-in migrant care workers' individual value is reduced to their productivity. It affects both workers' and employers' perception of workers' bodies and health.

5.1 Work for a Better Life

Most interviewees said they had migrated in pursuit of a better life. "Better" not only equated with economic advantage but also referred to a more secure and stable life. They were working for the families they had left behind, and those families gave them needed support. Keyla recalled her decision to work overseas thus: "If I had not come to Taiwan, I might have gotten married at that time. I was the oldest in my family. At that time, I wanted to share the financial burden with my mother and help my younger siblings get a better education." Like most interviewees, Keyla was from Java, which has a long history of labor emigration as work there has long been scarce (Chan, 2017). The island has few jobs that pay a living wage and migrants felt they had no choice but to seek work elsewhere.

While Keyla had no children, most interviewees had left children behind in Indonesia. They saw migration as a way to provide their children with a promising future. Siti, for example, had a seven-year-old daughter:

I learn[ed] from my friend's experience. She worked in Taiwan to save a huge amount of money. She is a success now. She owns her business and has the ability to provide her family with a good life. My girl is still little now. It is the time I can leave her back home in my mother and younger sister's care. I have to work hard to save money. Later I can return home to be with her and prevent her from getting spoiled. I also have a dream. I hope that I can take care of my mother and my child in the future.

May was enduring abuse because she saw it as the only way to improve her children's lives. She had been a live-in care worker in Taiwan for eight years, having started at the age of 22. She said:

A-Ma has a bad temper. But I understand that it is her personality. I have to endure her. I didn't have another choice. … I would cry in the beginning. She beat me every day. She also bit me or squeezed me. … I couldn't stop crying. In the beginning, I was afraid of speaking to [the grandmother's adult child, her employer] about the A-Ma's situation. I told my employer that I felt helpless. I was thinking of running away but I could not. I had two children. If I did not [make money], they could not continue their education. My poor children! I had to cheer up no matter how bad the situation was.

Gender roles shape the migration experiences of this group of women (Ryan, 2008). Their identities as mother, daughter, and, for some, wife affect their decisions about working overseas and their work experiences in Taiwan. These experiences in turn affect their mental and physical health. On the one hand, emotional ties with their families back in their home countries give live-in migrant workers the spiritual support to sustain them in foreign lands; on the other hand, these ties have become trammels that constrain their agency. No matter how their employers or care recipients treat them, they have to endure.

5.2 "Our Health is not Their Concern": Disposable Workers

Many interviewees mentioned that they have no sick leave because of the characteristics of care work. Although all documented migrant workers in Taiwan are entitled to National Health Insurance, employers need to allow them to access it. And indeed, workers have to continue providing care even when they need care themselves. Most workers understand that for the employers their value is based on their workable bodies.

Puji had worked in Taiwan more than ten years. She described the difficulties live-in migrant care workers face thus:

> Some of my friends, they do not have money [for a co-pay to see a physician]. Often, neither their brokers [representatives of the recruiting agencies] nor their employers would care for them. The employers think that you are here to work. The elderly need someone to take care of them. If [a migrant worker] were sick, what could the employer do? Should she take care of A-Ma [herself]? [My] broker regards [me] as trouble [if I ask about sick time].

Puji mentioned that when she was sick she still had to work without exception. She felt that her employer regarded her as a machine rather than a human being. Siti was 36 years old and had been working overseas since she was 18. She said:

> I told my employer that I had a backache after enduring the pain for a while. But she [the daughter in-law of the elderly couple] said that it was normal because of the cold and humid weather during wintertime. She did not mention anything about seeing a doctor. She just suggested that I needed to have more rest.

Siti felt that her employer was too busy to bother with her back pain. She decided to keep silent until she could not endure the pain anymore. She mentioned that every day when she took a shower, she used hot water to relieve her pain. Many interviewees had back pain, and May was among the suffering. She needed to carry the old woman she cared for.

She had been able to see a doctor, but she had to depend on shots of anodyne in order to work. Live-in migrant care workers who suffered from chronic pain had no way to recover as their job continually required them to repeat the movements that had injured them.

Wati had seen a psychiatrist for more than six months after the woman she took care of died and she herself had been diagnosed with post-traumatic stress disorder, but she still had to care for the woman's widower. She had problems sleeping and was afraid of staying with him alone when it got dark. But when I asked her about her health, she responded: "I am healthy. I do not have time to be sick." Wati told her employer (the cared-for's adult daughter) and the employment agency her symptoms at the beginning of her mental suffering. But neither had helped her see a psychiatrist. She had gotten care because a missionary and I accompanied her to see a doctor in turn. Most interviewees had found their employers displayed little concern for their health problems, and their monthly fees to their employment agencies did not cover health services. Workers generally recognized that they must maintain a workable body or be replaced.

6 Conclusion

This chapter analyzed the lives of live-in migrant care workers who are situated within particular social and political–economic contexts that constrain their agency. I utilized the concept of structural violence to illustrate the linkage between macro forces and individuals' daily experiences. This approach challenges the mainstream public health studies that focus on individual behavior and its relation to health risk. At the same time, it broadens the definition of violence to highlight both visible and invisible forces that affect the everyday lives of live-in migrant care workers. The findings draw our attention to social infrastructures and institutional arrangements, including state policy and regulations, liminality of migratory status, and power relations between workers, employers, and the cared-for that shape live-in migrant care workers' work conditions, health experiences, and well-being. The characteristics of care work and work in private households make their situation more vulnerable.

Most research participants suffered from the occupational hazards of care work, such as back pain and muscle strain. However, I focused on their work-related stress and mental health. I illustrated that migrant live-in care workers' motivation for pursuing better lives for their families is a double-edged sword. On the one hand, the families they leave behind provide them with support; on the other hand, workers can be trapped in terrible work conditions because they prioritize benefits to their families as mothers, daughters, sisters, or wives. Many of them describe their lives as filled with stresses and exhaustion that result from long work hours, heavy care responsibilities, and the dynamics of care provision. The employers of live-in migrant care workers and the brokers usually regard them as workers rather than full human beings. The workers' value is judged by their workable bodies and the labor they provide.

The evidence presented here suggests the cost when the government refuses its critical role of regulating working conditions in the context of globalization and transnational migration. To date Taiwan has focused on regulating live-in migrant workers rather than providing them with legal protection and social support. This has left workers highly vulnerable. The Labor Standards Act should cover live-in migrant care workers, whose right to mental and physical health is being violated in the absence of state regulation. Beyond this, truly protecting live-in workers' rights will involve rethinking the logic of capital production that works together with a patriarchal system and racialized ideology, devaluing the care work performed by women from the so-called Third World countries.

Notes

1. https://law.moj.gov.tw/ENG/LawClass/LawAll.aspx?pcode=N0090028
2. The amount of the service fee should not be over NT$1800 for each month in the first year, NT$1700 for each month in the second year and should be not be over NT$1500 for each month after the third year according to the foreigner working cumulative time into the country at the time. However, the fee should not be over NT$1500 for each month if the employee has been hired for more than two years, or the employee has to leave the county and make re-entry due to the termination of

employment or the expiration of employment duration and is hired again by the same employer.

3. https://law.moj.gov.tw/ENG/LawClass/LawAll.aspx?pcode=N0090002

References

Ahonen, E. Q., Lopez-Jacob, M. J., Vazquez, M. L., Porthe, V., Gil-Gonzalez, D., Garca, A. M., et al. (2010). Invisible work, unseen hazards: The health of women immigrant household service workers in Spain. *American Journal of Industrial Medicine, 53*(4), 405–416.

Anderson, B. (2000). *Doing the dirty work: The global politics of domestic labour*. Zed Books.

Anderson, B. (2007). A very private business: Exploring the demand for migrant domestic workers. *European Journal of Women's Studies, 14*(3), 247–264.

Arcury, T. A., & Quandt, S. A. (2007). Delivery of health services to migrant and seasonal farmworkers. *Annual Review of Public Health, 28*, 345–363.

Atanackovic, J., & Bourgeault, I. L. (2014). *Economic and social integration of immigrant live-in caregivers in Canada*. IRPP.

Ayalon, L. (2009). Family and family-like interactions in households with round-the-clock paid foreign carers in Israel. *Ageing and Society, 29*(5), 671–688.

Benson, P. (2008). El campo: Faciality and structural violence in farm labour camps. *Cultural Anthropology, 23*(4), 589–629.

Bialon, L. N., & Coke, S. (2012). A study on caregiver burden: Stressors, challenges, and possible solutions. *American Journal of Hospice & Palliative Care, 29*(3), 201–218.

Brotheridge, C., & Grandey, A. (2002). Emotional labour and burnout: Comparing two perspectives of 'people work'. *Journal of Vocational Behaviour, 60*, 17–39.

Chan, C. (2017). Not always 'left-behind': Indonesian adolescent women negotiating transnational mobility, filial piety and care. *The Asia Pacific Journal of Anthropology, 18*(3), 246–263.

Chang, G. (2016). *Disposable domestics: Immigrant women workers in the global economy*. Haymarket Books.

Cheng, S. J. A. (2003). Rethinking the globalization of domestic service: Foreign domestics, state control, and the politics of identity in Taiwan. *Gender & Society, 17*(2), 166–186.

Da Conceição Figueiredo, M., Suleman, F., & do Carmo Botelho, M. (2018). Workplace abuse and harassment: The vulnerability of informal and migrant domestic workers in Portugal. *Social Policy and Society, 17*(1), 65–85.

Datta, K., McIlwaine, C., Evans, Y., Herbert, J., May, J., & Wills, J. (2006). *Work, care and life among low-paid migrant workers in London: Towards a migrant ethic of care*. Working Paper. Department of Geography, Queen Mary, University of London.

Ehrenreich, B. (2002). Maid to order. In B. Ehrenreich & A. R. Hochschild (Eds.), *Global woman: Nannies, maids, and sex workers in the new economy* (pp. 85–103). Henry Holt.

Farmer, P. (2004). *Pathologies of power: Health, human rights, and the new war on the poor*. University of California Press.

Frantz, E. (2008). Of maids and madams. *Critical Asian Studies, 40*(4), 609–638.

Galtung, J. (1969). Violence, peace, and peace research. *Journal of Peace Research, 6*(3), 167–191.

Galtung, J. (1975). *Peace: Research, education, action*. Christian Ejlers.

Holmes, S. M. (2007). 'Oaxacans like to work bent over': The naturalization of social suffering among berry farm workers. *International Migration, 45*(3), 39–68.

Holmes, S. M. (2013). *Fresh fruit, broken bodies: Migrant farmworkers in the United States*. University of California Press.

Huang, S., & Yeoh, B. S. (2003). The difference gender makes: State policy and contract migrant workers in Singapore. *Asian and Pacific Migration Journal, 12*(1–2), 75–97.

Khan, S. A. (2009). From labour of love to decent work: Protecting the human rights of migrant caregivers in Canada. *Canadian Journal of Law & Society/La Revue Canadienne Droit et Société, 24*(1), 23–45.

Lan, P. C. (2006). *Global Cinderellas: Migrant domestics and newly rich employers in Taiwan*. Duke University Press.

Liang, L. F. (2011). The making of an 'ideal' live-in migrant care worker: Recruiting, training, matching and disciplining. *Ethnic and Racial Studies, 34*(11), 1815–1834.

Liang, L. F. (2014). Live-in migrant care workers in Taiwan: The debate on the Household Service Act. *Asian and Pacific Migration Journal, 23*(2), 229–241.

Liang, L. F. (2015). The distortions of care needs and medical professionalism: The ruling practices of migrant labor policy in Taiwan. *Journal of Aging Studies, 34*, 29–37.

Liang, L. F. (2017). Migrant live-in care workers in Taiwan: Multiple roles, cultural functions, and the new division of care labour. In K. Christensen & D. Pilling (Eds.), *The Routledge handbook of social care work around the world* (pp. 228–240). Routledge.

Liang, L. F. (2018). Migrant care work in Taiwan: Applying a feminist ethics of care to the relationships of 'live-in' care. *International Journal of Care and Caring, 2*(2), 215–233.

Lin, S., & Bélanger, D. (2012). Negotiating the social family: Migrant live-in elder care-workers in Taiwan. *Asian Journal of Social Science, 40*(3), 295–320.

Lindio-McGovern, L. (2003). Labour export in the context of globalization: The experience of Filipino domestic workers in Rome. *International Sociology, 18*(3), 513–534.

Lutz, H. (2011). *The new maids*. Zed Books.

Mayell, S., & McLaughlin, J. (2016). Migrating to work at what cost? The cumulative health consequences of contemporary labour migration. In F. Thomas (Ed.), *Handbook of migration and health* (pp. 230–252). Edward Elgar Publishing.

Miller, M. J., & Castles, S. (2009). *The age of migration: International population movements in the modern world*. Palgrave Macmillan.

Ministry of Labor. (2020a). Foreign workers statistics. Retrieved June 21, 2020, from http://statdb.mol.gov.tw/html/mon/c12010.htm

Ministry of Labor. (2020b). The report of application and management of foreign workers in Taiwan 2019. https://english.mol.gov.tw/homeinfo/6458/6555/6566/

Ministry of the Interior. (2020). Statistical yearbook of Interior. Retrieved March 15, 2020, from https://www.moi.gov.tw/files/site_stuff/321/2/year/year_en.html

Näre, L. (2011). The moral economy of domestic and care labour: Migrant workers in Naples, Italy. *Sociology, 45*(3), 396–412.

National Development Council. (2018). Population project for Taiwan: 2018–2060. Retrieved March 15, 2020, from http://www.ndc.gov.tw/en/cp.aspx?n=2E5DCB04C64512CC

Paradise, M., McCade, D., Hickie, I. B., Diamond, K., Lewis, S. J., & Naismith, S. L. (2015). Caregiver burden in mild cognitive impairment. *Aging & Mental Health, 19*(1), 72–78.

Popova, N., & Özel, M. H. (2018). *ILO global estimates on international migrant workers: Results and methodology*. International Labour Office.

Porat, I., & Iecovich, E. (2010). Relationships between elderly care recipients and their migrant live-in home care workers in Israel. *Home Health Care Services Quarterly, 29*(1), 1–21.

Pratt, G. (2012). *Families apart.* University of Minnesota Press.

Ryan, L. (2008). 'I had a sister in England': Family-led migration, social networks and Irish nurses. *Journal of Ethnic and Migration Studies, 34*(3), 453–470.

Silva-Smith, A. L. (2007). Restructuring life: Preparing for and beginning a new caregiving role. *Journal of Family Nursing, 13*(1), 99–116.

Stevens, M., Hussein, S., & Manthorpe, J. (2012). Experiences of racism and discrimination among migrant care workers in England: Findings from a mixed-methods research project. *Ethnic and Racial Studies, 35*(2), 259–280.

Teshuva, K., Cohen-Mansfield, J., Iecovich, E., & Golander, H. (2019). Like one of the family? Understanding relationships between migrant live-in care workers and older care recipients in Israel. *Ageing & Society, 39*(7), 1387–1408.

Timonen, V., & Doyle, M. (2010). Migrant care workers' relationships with care recipients, colleagues and employers. *European Journal of Women's Studies, 17*(1), 25–41.

Walsh, K., & O'Shea, E. (2010). Marginalised care: Migrant workers caring for older people in Ireland. *Journal of Population Ageing, 3*(1–2), 17–37.

Walter, N., Bourgois, P., & Loinaz, H. M. (2004). Masculinity and undocumented labour migration: Injured Latino day laborers in San Francisco. *Social Science & Medicine, 59*(6), 1159–1168.

Xiang, B., & Lindquist, J. (2014). Migration infrastructure. *International Migration Review, 48*(1_suppl), 122–148.

Yeoh, B., & Huang, S. (2010). Foreign domestic workers and home based care for the elderly in Singapore. *Journal of Aging & Social Policy, 22*, 69–88.

11

Transnational Social Protection for Migrant Care Workers: The Experiences, Practices of and Hurdles for Self-Employed 24-Hour Care Workers

August Österle and Gudrun Bauer

1 Introduction

Migrant care work in private households has developed into a major pillar of many countries' long-term care systems around the world (Christensen & Pilling, 2018). In the European context, mostly women from Central and Eastern European countries are traveling to Western and Southern European countries to provide help and support to older people in need of long-term care, either as live-ins staying in the home of the user or as live-outs. However, the underlying arrangements differ widely (e.g., Da Roit & Weicht, 2013; van Hooren, 2013; Williams, 2012). In many countries, migrant care work in private households emerged as a gray economy of care with irregular work arrangements.

A. Österle (✉) • G. Bauer
WU Vienna University of Economics and Business, Vienna, Austria
e-mail: august.oesterle@wu.ac.at

V. Horn et al. (eds.), *The Global Old Age Care Industry*,
https://doi.org/10.1007/978-981-16-2237-3_11

Several countries have seen efforts of regularization, but often illegal and legal arrangements co-exist. These differences in work arrangements also have huge implications for formal social security coverage of migrant care workers. In addition, the particular pattern of work migration—often rotational migration—has further implications for whether, where (in the home or in the destination country) and to what extent workers have access to social security rights.

This chapter explores the experiences and practices of intra-EU migrant care workers with formal social protection. More specifically, it studies whether and how self-employed migrant care workers make use of the provisions determined by EU social security coordination and national regulations. The analysis focuses on Austria as destination and Central and Eastern European countries as countries of origin. Austria stands out in two ways: first, the regularization of previously illegal[1] employment arrangements for domestic migrant care work has been most comprehensive in a European comparative perspective (Österle & Bauer, 2016). Second, arrangements build almost exclusively on self-employment of care workers (Österle, 2018). The study is situated in the European social citizenship literature. In theorizing European social citizenship, Amelina et al. (2020) differentiates three analytical dimensions of that concept: namely, rights and obligations, the rhetoric of legitimation and the participatory dimension. This study is particularly interested in linking the rights and obligations dimension of intra-EU migration with the participatory dimension—that is, with the lived experiences and practices of Central and Eastern EU citizens in accessing and using transnational social rights. It adds to the literature by focusing on a group of workers that combine several particular features (which distinguishes them from the majority of intra-EU migrants): the workers are self-employed; there is a rather strict division between the place of work in one country and the place of stay while not working in another country; the arrangements are based on rotational migration with periods of "24-hour work" concentrated in two or more weeks followed by the same duration out of work in the home country.

Building on a review of the institutional context of national and transnational social protection in the European Union, the chapter uses information from in-depth interviews with 23 self-employed migrant care workers in Austria originating from Slovakia and Romania. After this introduction, Sect. 2 very briefly reviews the concept of transnational social protection and the major dimensions relevant for the study of (formal) transnational social protection for migrant care workers. Section 3 then introduces the context for migrant care work in Europe, and in Austria in more detail, looking at the type and the features of migrant care work arrangements and the regulatory context for social security coverage. The key results on practices and experiences with transnational social protection coverage are then presented and discussed in Sect. 4. In Sect. 5, a brief summary and outlook will conclude the chapter.

2 Transnational Social Protection

Research on the transnational social question and on transnational social policy, social protection and social support has grown substantially in the past decade: in a global perspective (e.g., Faist 2019; Chambon et al., 2012; Sabates-Wheeler & Feldman, 2011), with a view to the specific context of European social security coordination (e.g., Amelina et al., 2020; Pennings & Seeleib-Kaiser, 2018), and for the context of transnational aging (e.g., Horn & Schweppe, 2019). Migration for care work involves many of these transnational social questions, not least the question about social protection coverage. Care work migration—from the migrant worker perspective—is driven by the search for work and income, above all. This search for work and income is in fact a search for opportunities to improve the living conditions of oneself and those of family members. Traveling abroad for work can enhance opportunities to address social needs and social risks, such as ill-health, unemployment, otherwise unaffordable education for children or low income in old age. Cross--border mobility can create opportunities for formal and for informal social protection, largely depending on what migration and work regimes are available.

In reality, formal and informal social protection overlap and interact, and it is not just formal and/or informal social protection. In exploring the transnational social question, Faist (2019) depicts the assemblages of cross-border social protection by not only using the degree of formalization, but also scale. In this conceptualization, state-based schemes as well as bilateral or multilateral social security agreements are on the formal and systemic side. This particular aspect will be at the center of this analysis. But formalized, systemic social protection is not independent of other forms of social protection. Private market solutions, for example, are another formalized way of social protection, but one that is usually less or not accessible to those in low-income jobs. NGOs may take very different roles, providing state-funded provisions, providing information and support independent from state funding or acting in an advocacy role. Social networks are vital for information as well as social and emotional support. And, of course, families have a key role in formal and informal social protection. In the case of migrant care workers, family members are not only recipients of income earned abroad, but also potential addressees of formal social rights (e.g., family benefits), recipients of informal social protection or members of family networks in which, for example, care-giving roles are re-arranged when one family member is not available due to work abroad.

This contribution explores the practices and experiences of intra-EU migrant care workers with formal social protection. From a global perspective, irregular arrangements or highly precarious semi-legal arrangements dominate migrant care work or migrant domestic work in private households. Informal transnational social protection is therefore much more substantive in these cases than formal social rights (Boccagni, 2017). However, there are examples where formal and portable social rights are established. In the EU context, care work arrangements in the private household that comply with labor and social security regulations establish transnational social rights for migrant care workers as EU citizens. The following sections will explore the context of these rights and the experiences of migrant care workers in exercising these social rights.

3 Migrant Long-Term Care Work and Social Protection

In several European countries, mostly in Southern Europe and in the German-speaking countries, migrant care workers in private households have become a major source of long-term care provision for older people in need of help and support. In Southern European countries migrant care work became increasingly important from the mid-1990s, at that time with large numbers of care workers originating from Latin America. From the 2000s, the proportion of migrant care workers originating from Central and Eastern European countries increased sharply (Da Roit et al., 2013). In Italy, Pasquinelli (2013, as cited in Di Rosa et al., 2018) estimated the number of migrant care workers for the early 2010s to reach more than 830,000 people, including those with and those without a regular contract. In the German-speaking countries, migrant care work in private households became a novel but irregular care work arrangement from the late 1990s and more significantly from the early 2000s. In these countries, care workers were and still are almost exclusively from Central and Eastern European EU member countries. For Germany, Satola and Schywalski (2016) give an estimate of between 300,000 and 400,000 migrant care workers. In Austria, by the end of 2019, 62,000 individuals were registered as active self-employed care workers (WKO, 2020).

3.1 The Variety of Migrant Care Work Arrangements

The most important common feature of migrant care work in private households is the provision of care, help, support and attendance for those in need of long-term care in their private homes. The concrete arrangements, however, differ widely. While live-in and live-out arrangements co-exist in Southern European countries, the live-in arrangement dominates in the German-speaking countries. While respective arrangements mostly developed as a gray economy of care, many countries have seen more or less successful efforts of regularization, often a series of regularizations (see, e.g., Cordini and Ranci, 2016 for the Italian case). Today, in most countries, regular and irregular work arrangements co--exist. In Germany, for example, regular work arrangements in private households predominantly follow the posted worker model, with care workers employed according to the EU-posted workers' directive and placed by their employer to carry out services in another EU member state. At the same time, and above all for reasons of costs and flexibility, irregular arrangements remain important in Germany (Theobald, 2018). However, even if regular work arrangements are established, these are often highly precarious work arrangements.

Austria stands out in terms of regularization, as it has been most successful in terms of the take-up of the regular arrangement. This is largely because of the self-employment mode that was made possible with the reform in 2006–2007 (Österle & Bauer, 2016). In 2006, right before national elections, the illegality of migrant care work arrangements became a huge political issue. It started with charges filed by social insurance institutions for not paying contributions. However, it was the illegality of a service that was widely valued which fueled the debates. Soon, all major political parties agreed on regularization that should allow households to continue with the existing arrangement of two or more weekly shifts of migrant care workers in the private household and without substantial additional costs to the household. Even though the law provides the opportunity for employment and self-employment of 24-hour care workers, self-employment became the near-exclusive arrangement. This is largely for economic reasons and due to the flexibility that self-employment provides. While experts have repeatedly questioned whether migrant care work in fact is self-employment (e.g., Schaup, 2020), allowing migrant care work on a self-employment basis made it possible to continue with highly flexible work time arrangements. It was possible to continue with a two or more weekly shift arrangement (rotating between work stays in Austria and stays in the home country); and it allowed the self-employed to work very long hours, a constellation that would not be possible under an employer-employee arrangement. Additionally, the costs of the regular arrangement did not increase substantially beyond the cost of an irregular arrangement. This is largely because work time regulations, minimum pay regulations and regulations on vacation and sick leave for employees do not apply for the self-employed. However, social security contributions and taxes (in the latter case beyond a threshold that migrant care workers usually do not reach) do apply. To contribute to the coverage of these additional costs, and to lower the extra costs arising with regularization for private households, Austria has introduced a means-tested benefit for migrant care work arrangements. Apart from the means testing, the benefit requires that the care provided is for someone in benefit level 3 or above[2] and that the worker meets minimal qualification requirements[3] (Österle, 2018). Placement of migrant care workers is organized by agencies, mostly sole traders or private limited companies. They differ considerably in their basic characteristics and the services they provide (Leiber et al., 2021). By the end of 2019, 826 placement agencies were registered (WKO, 2020).

In terms of migration patterns, migrant care workers in private households originating from Central and Eastern Europe are moving between countries on a rotational basis. This is typical for live-in migrant care workers, the dominant mode in the German-speaking countries but also widespread in Southern Europe. In the live-in arrangement, migrant care workers are living in the home of the user, with board and lodging as part of the financial arrangement. The rotational migration pattern implies that the country of work is different from the country where care workers live when not working. In most cases, family members of the care workers stay in their home countries. This has several implications: for the arrangement of work and family in two different countries, for traveling between these two spaces, for providing paid care in one country while often also providing informal unpaid care back home (e.g., Bauer & Österle, 2016). Moreover, it has huge implications for migrant care workers' access to social rights. European rules for social security coordination (Regulations 883/2004 and 987/2009) and national rules of the destination and of the source country have an impact on social security coverage. Hence, social rights will be determined by the interplay of European, national and partly even sub-national regulations and the choices individuals make with a view to what they know about their rights, how they view their rights and their abilities to materialize their rights.

3.2 Formal Social Security for Migrant Care Workers

Given the complexity and variety of migrant care work arrangements, the realities of social protection for migrant workers vary widely. In the case of a regular work arrangement, workers have access to formal social protection (which, however, may be limited depending on the specific features of the particular arrangement). If work arrangements do not comply with labor and social security regulations (as is the case for many migrant care workers in private households in Southern European countries but also in Germany), workers may still have access to formal social protection via other routes, for example, via family or residence status in their home country, but not via the paid care work they are providing. The lower the level of formal state-based social protection, the more care workers depend on alternative provisions.

For formal social security coverage of migrant care workers in private households in Austria, three features are important: First, migrant care workers in private households in Austria are almost exclusively from Central and Eastern European Union member states. Above all, it is workers from Romania and Slovakia. These two countries account for more than 80% of all self-employed care workers in Austria (Österle, 2018). For all these workers, European social security coordination applies. Second, they migrate on a rotational basis between the country of origin (and stay if not working in the Austrian household) in Central and Eastern Europe and the country of work, Austria. The rotational mode was always perceived as convenient to combine working in Austria with higher earnings and living in the home country (Österle & Bauer, 2016), but it is not defined as a legal requirement for the job in Austria. Third, migrant care workers in private households in Austria are self--employed, implying that the rules that apply may differ from the rules that apply for employees.

The comparative welfare state literature describes Austria as a Bismarckian or as a corporatist-conservative welfare state (for an overview see Österle & Heitzmann, 2020). Differentiation in social policies according to occupational groups is a key feature of Bismarck-oriented welfare states and of the development and of the status-quo of social policies in Austria. Social insurance is the predominant arrangement for four main welfare state pillars, namely pensions, health care, accidents and old age, while family policies and long-term care policies are universally oriented, tax-funded welfare state sectors. In these two sectors, cash benefits have a comparatively larger role than services. Historically, and in many respects up until today, social insurance programs differentiate between occupational groups, in particular between employees and the self-employed; and family status plays a more dominant role than, for example, the Northern European welfare state model.

Below, the potentially main social programs and the regulations for intra-EU migrant care workers are briefly outlined without going into the legal details.[4] The analysis of practices and experiences of migrant workers with regard to social protection in this chapter refers to health care, unemployment, family benefits and pension rights.

Health care coverage in Austria is organized in several sickness funds. From 2020, previously nine regional sickness funds (mainly for workers) are merged into one Austrian health fund, while the sickness fund for the self-employed is merged with the sickness fund for farmers. Hence, a long-debated merging of funds followed the traditional division between occupational groups. While earlier reforms have harmonized contribution rates, benefit packages and co-payments still differ between funds. In the current situation, differences for the self-employed—compared to employees—exist in terms of benefit packages, but more importantly, they have to make substantial co-payments for services used. In the frame of European social security coordination, migrant care workers are paying contributions to social health insurance in Austria and they are entitled to health care in Austria. Because of the co-insurance principle in the Austrian health care system, these entitlements also include dependent children.

Historically, the *pension* system was also organized according to occupational background. With a major pension reform in 2004–2005, a process of harmonization has begun. Self-employed migrant care workers pay contributions to the Austrian social pension fund. In the European context, for pensions, the accumulation of months of insurance is of particular importance. According to Austrian legislation, the entitlement to an old-age pension requires that the person can prove at least 180 months of insurance in Austria or any other EU member state (following the principle of accumulation of insurance months). If a migrant care worker has been paying pension insurance contributions for at least one year in Austria and if the accumulated months of insurance in the EU are proved, (s)he establishes a right to receive an Austrian public old-age pension. With a view to the low levels of old-age pensions in countries like Romania, an Austrian pension right can become very significant but is of course highly dependent on the number of years care workers have been paying contributions to the Austrian pension system.

A major difference between employees and the self-employed exists for *unemployment*. Unemployment insurance is not mandatory for the self--employed in Austria. The possibility to voluntarily opt into unemployment insurance—and to choose one of three different levels of contributions and benefits—was only introduced in 2009 for this group. According to European social security coordination, portability of unemployment benefits is possible after an application. Hence, migrant care workers out of work in Austria—in case they have opted into voluntary unemployment insurance and in case they fulfill the eligibility requirements in terms of wait period—could apply to transfer the benefit to their respective home country. This, however, requires that they decide to look for work there and to follow the respective requirements of that country. As with employees, social protection of the self-employed for sickness, old-age pensions and unemployment follows social insurance principles. However, there is an additional specific feature for the self--employed. For the first three years, contributions for health and pension insurance can be reduced to facilitate and support entrepreneurial activities.

The fourth main element of the benefit system is *family benefits*. Different from the aforementioned social rights, these are universal and tax-funded benefits. Family benefits range between €114.00 and €165.10 per child and per month (in 2020), depending on the age of the child, and they further increase for siblings. Family benefits are paid as long as the child is in education, including tertiary education, potentially up until age 24. According to European coordination rules, family benefits follow priority rules if more than one EU member state is involved. In general, family benefits are paid by the country in which a parent is working and where the family lives. When both parents work in different countries (e.g., parent A in Austria, parent B in Slovakia) but the children's main residence is in one of the countries (e.g., in Slovakia), then the latter country is responsible for paying the family benefits. However, if the amount of the family benefit is lower in the competent state (e.g., in Slovakia) than in the other country (e.g., in Austria), then the country with the higher benefit (e.g., Austria) has to top up the level of the lower benefit in the other country (e.g., in Slovakia).

In 2018, eligibility for family benefits became a hugely controversial issue in relation to migration and European social security coordination in Austria (but also other European countries). The then center-right coalition government pushed for a reform, which finally led to a major change from January 2019. Accordingly, Austria introduced an indexation of family benefits. Benefits are adjusted according to the living costs in the respective country. As a consequence, the family benefit for a child living in Slovakia is less than two-third and that for a child living in Romania is less than half the amount of that for a child living in Austria. There is broad but not unanimous consensus that this is in conflict with European rules. Consequently, in 2019, the European Commission opened an infringement procedure against Austria. As the case was not settled, in May 2020, the Commission referred the matter to the European Court of Justice.

Apart from access to benefits in Austria (health care) and apart from the portability of Austrian social security benefits (family benefits, pension rights, unemployment benefits), migrant care workers may also have access to respective provisions in their home country. As just outlined, this applies to family benefits (the country of work topping up respective levels), but may also apply to health care (in the case of residence-based coverage or coverage as a family member) or pensions (in case of previous years of work and contribution in the respective country).

4 Experiences and Practices of Self-Employed Migrant Care Workers

This section, building on the regulatory context for transnational social protection and the particular migrant care work arrangement in Austria, explores experiences and practices of Central and Eastern European migrant care workers in Austria. The analysis builds on in-depth interviews with 23 self-employed migrant care workers originating from Slovakia (13) and Romania (10). The interviews were part of a larger study dealing with the transnational situation of live-in care workers from Central and Eastern European countries in Austria. In particular, the study focused on transnational formal and informal obligations of female care workers, work experiences and experiences of informal and formal social support. Placement agencies were helpful in accessing the field, while the snowball method was followed after the first few interviews. At the time of the interview, most interviewees were aged between 40 and 60 years, which is highly representative for these workers. They were all self-employed care workers in a private household in Austria. The selection of interviewees specifically tried to reach workers with dependent children, hence a population that potentially is eligible for family benefits. The interviews addressed experiences with social protection coverage in general as well as specific elements of coverage, most importantly family benefits, health care coverage, pension rights and unemployment benefits. The analysis of the material followed the framework analysis method (Ritchie et al., 2003).

Overall, the patterns emerging from the interviews show considerable variations in the practices and experiences of migrant care workers. Partly this relates to the take-up of benefits. While potential benefits from pension insurance and unemployment insurance lie in the future, health care and child care benefits are potentially relevant at the same time as paying contributions. The following analysis will firstly present the variations; it will then explore key factors determining the use of social rights attached to self-employment in Austria (information, migration patterns, cost and quality of provisions and personal circumstances) and will finally discuss the interplay of contextual factors on macro, meso and micro levels.

4.1 Huge Varieties in Experiences, Practices and Perceptions

The findings indicate considerable variety in the practices, the experiences and the perceptions toward the transnational take-up of social benefits. Above all, the overall perception of having social security coverage ranges widely. Although some care workers are aware of the more extended social protection that comes with an employer-employee contract, self-employment based on the rotational system is perceived as more convenient—allowing rotational migration—and thus widely preferred to contract employment in Austria (Österle & Bauer, 2016). For the majority of interviewees, social rights are a major additional benefit of having a regular work arrangement. Some interviewees even identify this as a co-driver of care work migration, in particular when they talk about old-age security (pointing at the need to top up very low pensions and to gain pension rights). Other interviewees, instead, also emphasize the additional financial burden social insurance contributions create. They argue that they are paying a lot into the system while getting very little out of it, assuming an imbalance between obligations to make contributions and rights to receive support. A few even argue that social insurance contributions lower their income. They would therefore prefer an arrangement outside the regular system (cf. Menebröcker in this volume). A Slovakian care worker who has also experienced work in an irregular arrangement addresses both sides:

> *Official [regular employment] is good on the one hand. But this system in Austria is kind of hard for us. We really pay a lot into the system and we get almost nothing in return. … Better if I work without papers.* (Slovakian, 61 years)

Respective assessments not least result from vast differences in individual circumstances and from whether care workers are currently using specific benefits (most importantly health care and family benefits). Health care use in Austria, for example, differs widely between care workers. Family benefits are relevant for those with children but not for others. The use of pension rights lies far in the future for some, while it was a push factor for going abroad for work for others. In this context, it is not surprising that some are regarding social insurance coverage in Austria as an additional cost rather than as a potential benefit. Others, instead, emphasize and value the benefits they are receiving in health care or for child care while working, or what they expect to receive in the future from the Austrian pension system. These variations in perceptions also translate into the experiences and practices care workers report, which will be discussed in the next steps.

4.2 Information

First of all, the interviews indicate huge differences in the extent to which care workers are informed about their rights. A lack of information is not only an issue for those who have started migrant care work more recently. Some do not have to care about, for example, health provisions, because respective needs do not occur. Or, they only start thinking about the details of receiving an Austrian old-age pension when coming closer to retirement age.

> *Yes yes. I need to stay for one year, when taking a pension from you. I don't know, I don't get it.* (Romanian, 58 years)

Another factor is that administrative tasks are often taken over by the placement agencies (Leiber et al., 2021; Steiner et al., 2019). Most agencies in Austria act as intermediaries on a for-profit basis; few are attached to non-profit social services providers. The agencies are in charge of searching for care workers, assessing the needs and demands of care users and their families and placing 24-hour care workers. In reality, there are huge differences in the practices of the agencies. This applies to the financial arrangements. They can charge both users and care workers, and they are free in applying a lump sum fee or regular payments. Moreover, practices differ in terms of the administrative tasks they take over, despite the fact that self-employment usually implies that the self-employed themselves are in charge of certain administrative tasks. This, not least, is one of the reasons why self-employment of care workers is questioned (Schaup, 2020). If agencies take over administrative tasks related to social security coverage, care workers often need not care much about paying social security contributions and have limited knowledge about the details of their social rights in Austria and transnational social rights. This, however, also increases dependency on the agencies. Alongside the agencies, networks of migrant care workers are the main source of information and advice, while clients or their families have not been mentioned as relevant in this regard by the interviewees. With regard to social networks, these are particularly important via social media and via personal contact while traveling between home and destination country. As a result, many care workers are rather well informed, at least about those rights that they can realize right away, in particular family benefits. In the case of health care, needs often occur immediately and require very quick responses. In these cases, agencies and other care workers they can easily contact via social media become important. For social risks and potential social rights that lie further in the future, agencies and exchange with other care workers are the main sources of information.

In terms of being well or less well informed, many workers mention bureaucratic hurdles, a lack of understanding for language reasons or nontransparent and complex procedures.

> *[…] I also have problems at home [in Slovakia] because our GP doesn't accept my insurance and everywhere I have to take millions of photocopies.* (Slovakian, 55 years)

> *Because I don't understand everything, what I have done here. There [in Romania] it is clearer for, what they do, examination, and.* (Romanian, 59 years)

The different levels of advice and information and the perception of knowledge and capability to deal with administrative hurdles translate into differences in the assessment and the practices regarding specific provisions. These practices are co-determined by several additional factors.

4.3 Rotational Migration and Space

Secondly, and in addition to information, experiences and practices are shaped by the specific patterns of rotational migration. The work arrangement of migrant care workers in private households implies that they spend two or more weeks in Austria, followed by a period of the same duration in the home country. The Austrian term used for these arrangements, 24-hour care, already indicates the huge workload during the two or more weekly shift in Austria. This leaves very little room for activities outside the household, including activities to gather information (cf. Van Bochove and Zur Kleinsmiede in this volume). Depending on where the care worker stays for work, visiting social security institutions or visiting a doctor can become even more challenging.

During the period care workers stay back home for two or more weeks, using services in Austria often could become very time consuming and costly because of the distance. Extending a stay in Austria outside the work shift usually will not be possible in the home of the user because the other live-in care worker arrives for the respective shift. Hence, it would require moving to a hotel room or other accommodation, or traveling between home country and Austria. For Slovakians and Romanians, the distance is different. However, given that most Slovakian migrant care workers are from Central and Eastern Slovakia, traveling still requires several hours. When health care needs emerge while staying in the respective home country, for most care workers it is not an option to travel to Austria, simply because of the distance. Even if they would prefer treatment in Austria, they opt for using health care in the respective home country.

> *Normally it would be better for me to see the doctor in Austria but it is not so easy with transport and everything. It takes me 10–15 hours by bus, that's not funny. Therefore it is better to get treated in Romania and I stay at home until I recover.* (Romanian, 42 years)

4.4 Cost, Quality and Level of Provisions

Third, practices and experiences of using or not using social provisions in Austria are shaped by cost, quality and level of provisions. Costs involve transaction costs, such as for information and traveling between spaces, but also direct costs. Such costs arise when services have to be paid out of pocket or when co-payments have to be made. Here, a particular feature of health insurance for the self-employed in Austria applies. While employees only pay a moderate flat rate co-payment, the co-payment for the self-employed amounts to 20% (or 10%, for health promotion). Therefore, the use of health services can become costlier in Austria than in the respective home country. Another specific case is dental care. Care workers argue that even though dental care is excluded from the benefit package in Slovakia, it can still be cheaper than in Austria (where more treatments are covered but requiring substantial co-payments).

The cost dimension also appears in another respect. As mentioned before, the self-employed can voluntarily opt into unemployment insurance. Many care workers describe the risk of losing work as the worst-case scenario that could lead to extreme poverty and above all have major negative consequences for their children. However, care workers do not want respective fears coming too close. In addition, they do not take up the option of voluntary unemployment insurance. The main reason given is the additional costs. They can choose among three models with a monthly contribution of €46.99, €187.95 or €281.93. In addition, care workers assume that placement agencies (whom they are often paying a regular fee) will always provide them with an opportunity for a job. Hence, agencies become some kind of an alternative unemployment insurer.

> *You always have to work, always. ... I have three children back home. I always have to work, no matter how the patients are.* (Romanian, 50 years)

Another aspect is perceived quality and trust. In health care, quality considerations are mainly expressed in terms of the reputation of the health care system and in terms of trust in health care. As a Romanian worker points out:

I do no longer trust in Romania, because the very, very good doctors have gone. (Romanian, 58 years)

Concerns of reputation and trust also apply to other benefits, for example pensions. Many care workers are not aware of the details of the pension system, but some of them had very strong feelings about pension rights. A few interviewees were rather pessimistic about future pension rights in general and those attached to their current work in Austria. In general, however, care workers regard the right to receive an Austrian pension as an important advantage of social security coverage in Austria. At the same time, some express disappointment about the level of the old-age pension they expect to receive after a few years of work in Austria, especially when comparing the contributions they make each month with the pension right they are gaining. Several interviewees refer to a rule of thumb obviously used by migrant care workers: €10 per year. A Slovakian care worker argues:

When working here for 17 years, I will get €170 as a public pension from Austria. … That is not that much. (Slovakian, 49 years)

4.5 Personal Circumstances

Fourth, of course, it is personal circumstances that impact the individual assessment of the relative importance of certain social rights. For example, for many older interviewees, pension rights are particularly important. Working with a regular work contract for many years as a migrant care worker in Western Europe can lead to a public pension that significantly exceeds pension rights in the respective home country.

When I am working here [in Austria] I hope, I am able to receive an old-age pension in Romania and I can have a better life. A little bit better than others, you know. (Romanian, 42 years)

Family benefits are highly relevant for those with children. For them, this benefit is a very substantial addition to their income. It is regarded as highly supportive in a situation where mothers have to leave their children back home. For many of them, taking the burden of work migration is very much driven by the wish to better support their children, to provide them with opportunities they would not have if none of their parents would stay in Western Europe for work and income. In these situations, receiving or not receiving a family benefit makes a substantial impact financially. A Romanian care worker compares the family benefit to the pension her father receives:

> *That is better than my fathers' old-age pension. It is an income.* (Romanian, 50 years)

4.6 The Interplay of Macro, Meso and Micro Levels

The variations in practices, experiences and perceptions that occurred in the interviews are driven by several factors on macro, meso and micro levels and the interplay of these factors. So far, the analysis focused more on individual conditions, the level of information of care workers, the specific migration patterns, costs and perceived quality of provisions and personal circumstances. But these are shaped by the broader regulatory framework and institutional context and by meso-level actors and activities—most importantly, the role of agencies.

The challenges and barriers for EU citizens to access transnational social rights have been discussed in previous studies (e.g., Amelina et al., 2020; Pennings & Seeleib-Kaiser, 2018; Scheibelhofer & Holzinger, 2018) and also apply to intra-EU migrant care workers if their contract follows labor and social security legislation of the respective country. However, working in one country and residing with family in another country and the self-employment mode of migrant care work arrangements in the Austrian case can create additional challenges and limitations. In addition, the reality of 24-hour care work in private households poses considerable challenges for the practical realization of benefits. In these situations, the institutional meso-level context is of particular importance. As already discussed, agencies and networks of care workers are major sources of information, advice and support. In the interviews, care workers repeatedly refer to social networks of care workers, be it social media groups and communication or word-of-mouth on long bus trips between Romania or Slovakia and Austria. This becomes even more important when support by agencies is limited. Given the lack of regulations and monitoring for quality assurance in the brokerage market, the support that agencies provide varies widely (Leiber et al., 2021). Some agencies, in particular those linked to established welfare providers, have an important role here. They attempt to set standards beyond the legal requirements, to promote improvements in live-in migrant care work and to have an active role in keeping the public debate going about the risks and problems with 24-hour care work and brokerage. Finally, on the micro level it is not just exchange with other migrant care workers, but also the exchange with the care user and close family members. Again, the relationships, the understanding of the role as family member and/or employer and the role as care worker, and the support provided by these families vary considerably. Taken together, Austria—via the self-employment mode—was very successful in building a system for regular migrant care work in private households. However, the regulation is weak in terms of assuring quality in the brokerage process and in terms of work conditions (Leiber et al., 2021; Österle, 2018). This, not least, also leads to huge varieties in the practical challenges and experiences of realizing social rights.

5 Conclusion

Migrant care workers from Central and Eastern Europe staying in private households in Western and Southern Europe experience very specific work and migration patterns. Often, they are working outside labor and social security regulations, which put them in a particularly vulnerable situation. Others have regular work contracts but may still find themselves in precarious circumstances because of the low level of protection respective regulations provide or because of the intimacy of the private household as a workplace that might limit full compliance with the regulations. Austria in its regularization effort has chosen a particular arrangement, namely self-employment. This model of regularization has been very successful in terms of take-up of regular work arrangements. Hence, migrant care workers in Austrian private households benefit from the social security regulations attached to self-employment in Austria and, at least in principle, they can fully benefit from European social security coordination.

In reality, however, there are important limitations. The particular work arrangement (with a huge workload while in Austria) and the rotational migration pattern create limits to the practical choices workers can make. The complexity that can emerge in individual cases and a lack of knowledge and capacity to gather information can further limit the realization of rights. In addition, decisions to start searching for information and to try to materialize social rights are determined by individual situations and perspectives, and the importance ascribed to certain social benefits. In all this, social networks of migrant care workers have a key role in providing them with information and supporting them in the practical tasks needed for the take-up of benefits. The same applies to agencies, but their practical role as an effective source of information and support varies widely.

This chapter contributes to the European social citizenship literature (e.g., Amelina et al., 2020; Pennings & Seeleib-Kaiser, 2018) by linking the rights and obligations dimension of intra-EU migration with the practical hurdles for and the lived experiences of self-employed migrant care workers. In several European countries, migrant care work arrangements in private households are made outside labor and social security regulations, consequently also denying migrant care workers the benefits of EU social security coordination. The regularization of 24-hour care in Austria via self-employment has provided migrant care workers with the opportunities for transnational social protection and for better inclusion in European social citizenship. However, as shown in this contribution, challenges remain for using these opportunities, partly similar to those also known for other migrant workers in Europe, but partly very specific due to the rotational migration pattern, the 24-hour care arrangement, the self-employment mode and the weak regulation of work, work conditions and brokerage in this sector.

Notes

1. The illegality refers to the violation of labor law and social security law. Illegal and irregular are used synonymously in this chapter.
2. Austria grants cash-for-care payments to those in need of care. Those eligible are categorized between level 1 (the lowest level of care needs) and level 7 (the severest level of care needs), with benefits ranging between €160.10 and €1719.30 in 2020.
3. Qualification requirements can be fulfilled by proof of theoretical training equivalent to that of a home helper, or by proof of at least six months' experience in care for a person in need of care, or delegation of tasks by a medical doctor or a nurse.
4. For details, please refer to the literature (e.g., Pennings, 2015) and institutional websites providing details of a highly complex web of regulations on supranational, national and sub-national levels.

References

Amelina, A., Carmel, E., Runfors, A., & Scheibelhofer, E. (2020). *Boundaries of European social citizenship. EU citizens' transnational social security in regulations, discourses and experiences*. Routledge.

Bauer, G., & Österle, A. (2016). Mid and later life care work migration: Patterns of re-organising informal care obligations in Central and Eastern Europe. *Journal of Aging Studies, 37*(2), 81–93. https://doi.org/10.1016/j.jaging.2016.02.005

Boccagni, P. (2017). Addressing transnational needs through migration? An inquiry into the reach and consequences of migrants' social protection across borders. *Global Social Policy, 17*(2), 168–187. https://doi.org/10.1177/1468018116678523

Chambon, A., Schröer, W., & Schweppe, C. (Eds.). (2012). *Transnational social support*. Routledge.

Christensen, K., & Pilling, D. (Eds.). (2018). *The Routledge handbook of social care around the world*. Routledge.

Da Roit, B., González-Ferrer, A., & Moreno-Fuentes, F. J. (2013). The Southern European migrant-based care model: Long-term care and employment trajectories in Italy and Spain. *European Societies, 15*(4), 577–596. https://doi.org/10.1080/14616696.2013.836405

Da Roit, B., & Weicht, B. (2013). Migrant care work and care, migration and employment regimes: A fuzzy-set analysis. *Journal of European Social Policy, 23*(5), 469–486. https://doi.org/10.1177/0958928713499175

Di Rosa, M., Barbarella, F., Poli, A., Santini, S., & Lamura, G. (2018). Migrant workers in Italian households: Recent trends and future perspectives. In K. Christensen & D. Pilling (Eds.), *The Routledge Handbook of social care work around the world* (pp. 142–155). Routledge.

Faist, T. (2019). *The transnationalized social question. Migration and the politics of social inequalities in the twenty-first century*. Oxford University Press.

Horn, V., & Schweppe, C. (2019). Transnational mobilities of care in old age. *International Journal of Ageing and Later Life, 13*(2), 9–22. https://doi.org/10.3384/ijal.1652-8670.18-181227

Leiber, S., Rossow, V., Österle, A., & Frerk, T. (2021): Yet another black box: Brokering agencies in the evolving market for live-in migrant care work in Austria and Germany. *International Journal of Care and Caring, 5*(2), 187–08. https://doi.org/10.1332/239788220X15988973352874.

Österle, A. (2018). Employing migrant care workers for 24-hour care in private households in Austria: Benefits and risks for the long-term care system. In K. Christensen & D. Pilling (Eds.), *The Routledge handbook of social care work around the world* (pp. 130–141). Routledge.

Österle, A., & Bauer, G. (2016). The legalization of rotational 24-hour care work in Austria: Implications for migrant care workers. *Social Politics, 23*(2), 192–213. https://doi.org/10.1093/sp/jxv001

Österle, A., & Heitzmann, K. (2020). Austrification in welfare system change? An analysis of welfare system development in Austria between 1998 and 2018. In S. Blum, J. Kuhlmann, & K. Schubert (Eds.), *Routledge handbook of European welfare systems* (pp. 21–37). Routledge.

Pennings, F. (2015). *European social security law* (6th ed.). Intersentia.

Pennings, F., & Seeleib-Kaiser, M. (2018). *EU citizenship and social rights: Entitlements and impediments to accessing welfare*. Edward Elgar.

Ritchie, J., Spencer, L., & O'Connor, W. (2003). Carrying out qualitative analysis. In J. Ritchie & J. Lewis (Eds.), *Qualitative research practice: A guide for social science students and researchers* (pp. 219–262). SAGE.

Sabates-Wheeler, R., & Feldman, R. (Eds.). (2011). *Migration and social protection: Claiming social rights beyond borders*. Palgrave Macmillan.

Satola, A., & Schywalski, B. (2016). 'Live-in-Arrangements' in deutschen Haushalten: Zwischen arbeitsrechtlichen/-vertraglichen (Un-)Sicherheiten und Handlungsmöglichkeiten. In K. Jacobs, A. Kuhlmey, S. Greß, J. Klauber,

& A. Schwinger (Eds.), *Pflegereport 2016: Die Pflegenden im Fokus* (pp. 127–138). Schattauer.

Schaup, F. S. A. (2020). Rechtsfragen der 24-Stunden Betreuung. In W. J. Pfeil, A. Reichel, & S. Urnik (Eds.), *Pflege und Betreuung—Who cares?* (pp. 61–81). Manz.

Scheibelhofer, E., & Holzinger, C. (2018). 'Damn it, I am a miserable eastern European in the eyes of the administrator': EU migrants' experiences with (transnational) social security. *Social Inclusion, 6*(3), 201–209. https://doi.org/10.17645/si.v6i3.1477

Steiner, J., Prieler, V., Leiblfinger, M., & Benazha, A. (2019). Völlig legal!? Rechtliche Rahmung und Legalitätsnarrative in der 24h-Betreuung in Deutschland, Österreich und der Schweiz. *Österreichische Zeitschrift für Soziologie, 44*(1), 1–19. https://doi.org/10.1007/s11614-019-00337-4

Theobald, H. (2018). Care provision inside and outside the professional care system: The case of long-term care insurance in Germany. In K. Christensen & D. Pilling (Eds.), *The Routledge handbook of social care work around the world* (pp. 116–129). Routledge.

van Hooren, F. J. (2013). Varieties of migrant care work: Comparing patterns of migrant labour in social care. *Journal of European Social Policy, 22*(2), 133–147. https://doi.org/10.1177/0958928711433654

Williams, F. (2012). Converging variations in migrant care work in Europe. *Journal of European Social Policy, 22*(4), 363–376. https://doi.org/10.1177/0958928712449771

WKO. (2020). *WKO Statistik Österreich. Personenberatung und Personenbetreuung: Branchendaten*. Wirtschaftskammer Österreich.

12

Social Support Within and Outside Care Networks: Experiences of Live-in Migrant Care Workers in the Netherlands

Marianne van Bochove and Duco zur Kleinsmiede

1 Introduction

In the context of an aging population and revised long-term care policies, live-in migrant care (LIMC) has emerged as a niche market in the Netherlands (Da Roit & Van Bochove, 2017). Research in other national contexts has shown that live-in care often puts migrant care workers in precarious positions, in terms of both their labor market position and their position within the household of the care recipient (Hellgren, 2015;

This chapter draws on two earlier publications: Van Bochove and Zur Kleinsmiede (2020) and Van Bochove et al. (2017).

M. van Bochove (✉)
Governance of Urban Transitions Centre of Expertise, The Hague University of Applied Sciences, The Hague, The Netherlands
e-mail: m.e.vanbochove@hhs.nl

D. zur Kleinsmiede
VGZ, Utrecht, The Netherlands
e-mail: d.zurKleinsmiede@vgz.nl

V. Horn et al. (eds.), *The Global Old Age Care Industry*,
https://doi.org/10.1007/978-981-16-2237-3_12

Lutz, 2016; Salami et al., 2017). However, the outcome of precariousness is not self-evident. Salami et al. (2017) argue that national policies affect the circumstances of LIMC workers and that more research is needed on other underlying factors that influence the experience of precariousness. In this exploratory study, we provide more insight into such underlying factors, by focusing on the availability and nature of social support that LIMC workers in the Netherlands receive.

Drawing on literature about care networks (Jacobs et al., 2016; Broese van Groenou et al., 2016) and migrants' social networks (Ryan et al., 2008; Boccagni, 2014), we try to answer the following research questions: What role do various (client-related and non-client-related) actors play in LIMC workers' social support networks? How can differences in LIMC workers' access to emotional and practical support be explained, and what does this mean for their experience of precariousness?

In what follows, we will first discuss existing literature on the precarious nature of LIMC work and explain how our focus on care networks and migrants' social networks adds to this. Next, we describe important characteristics of LIMC work in the Dutch context and explain our research design. Then, we turn to our empirical research findings. We conclude with the implications of our findings for future research and policy.

2 Background

Live-in Migrant Care Work as Precarious Work

Live-in migrant care work is often studied from a precarious work perspective. In accordance with Kalleberg's (2009: 2) definition of precarious work, LIMC work is 'employment that is uncertain, unpredictable, and risky from the point of view of the worker'. The labor market position of LIMC workers is generally poor because of relatively low wages and limited social rights and protection (Lutz, 2016). On the household level, LIMC workers have to deal with unclear boundaries between work and leisure time and with unequal distributions of power between themselves and care recipients and their relatives. Salami et al. (2017: 1676) argue that treating the migrant caregiver as part of the family may offer support, but it can also be a strategy family members use 'to more easily manage or exploit' the caregiver. Isolation and strained relations with relatives 'back home' influence the psychosocial wellbeing of LIMC workers and may result in retreat as a coping strategy (Finch & Vega, 2003; Oxman-Martinez et al., 2004). LIMC work demands high levels of 'emotional labour' (Bauer & Österle, 2013), which can cause emotional exhaustion.

In this chapter, we are not so much interested in what defines precariousness, but rather in how precariousness is experienced by migrant workers themselves (cf. Wee et al., 2019). While Hellgren (2015) concludes that the position of migrant care workers is precarious regardless of national context, others have argued that governmental and organizational policies and practices seem to make a difference (Da Roit & Weicht, 2013; Leiber et al., 2021; Salami et al., 2017). In line with the approach of Wee et al. (2019), we argue that the degree to which LIMC workers experience precariousness depends on '*an interrelated number of conditions*' (p. 2677, italics in original). Access to social support can be one of these conditions. Earlier research showed that access to emotional support gained from social networks benefits the wellbeing of employees in emotionally challenging jobs (Kinman & Leggetter, 2016), but this has not been studied for LIMC work. Our combined focus on care networks and migrants' social networks provides insight into LIMC workers' access to various forms of social support and helps to explain why in the Netherlands some LIMC workers experience precariousness and others do not.

Social Support Within and Outside Care Networks

Based on literature on care networks and migrants' social networks, we argue that in studying LIMC workers' access to social support, it is important to pay attention to support *within* and *outside* care networks, *local* and *transnational* support, and different *forms* of support.

A care network perspective focuses on a collection of individuals who provide care for someone on a regular basis (Jacobs et al., 2016; Tonkens,

2012). Care networks can consist of formal caregivers, informal caregivers, or a mix of both (Broese van Groenou et al., 2016). Research on LIMC workers often focuses on dyad relationships between caregiver and care recipient or on a triad including the care recipients' children, mainly in their role as employers (Salami et al., 2017). However, LIMC workers often also have encounters with other paid caregivers in the care network, such as nurses, and with care recipients' children in their role as informal caregivers (Da Roit & Van Bochove, 2017). We assume that the composition and functioning of care networks can be important in shaping the experience of precariousness of LIMC workers. Client-related actors who provide different forms of support, not only directed at the care recipient but also at the LIMC worker, can alleviate LIMC workers' workload and enhance their emotional wellbeing. When such client-related actors are absent, do not provide support, or even hinder LIMC workers, LIMC workers are more likely to experience their working conditions as precarious.

Figure 12.1 gives an overview of the client-related actors included in this research.

While a focus on local care networks is important, it is not sufficient. Literature on migrants' social networks shows that most migrants do not only have social ties in the county of residence but also have transnational ties, such as with relatives and friends in the country of origin (Van

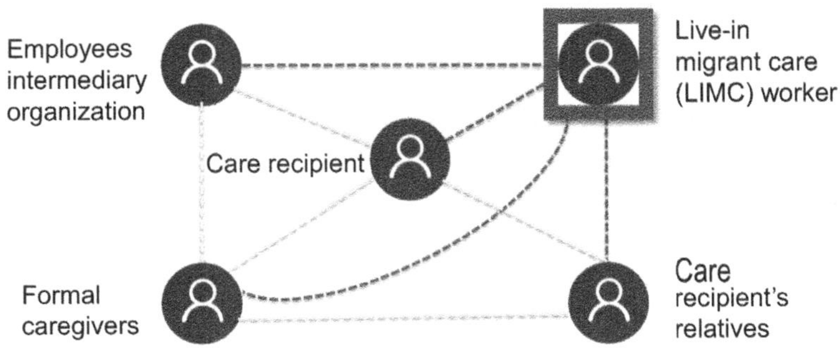

Fig. 12.1 Social support from actors within the care network

Bochove, 2012; Boccagni, 2014). For instance, Bruquetas-Callejo (2020) found that LIMC workers in the Netherlands often travel back and forth between the Netherlands and their home country. When they are in their home country, they do not have to work. Paradoxically, working abroad enables them to spend more time with their relatives. Ryan et al. (2008) argue that when studying migrants' social networks, both spatial and temporal aspects should be considered. Networks are not static but change with migrants' needs and circumstances. Newly arrived migrants, for instance, need different types of support networks than settled migrants.

In this study, we also explore the role of non-client-related actors—in other words, actors outside care networks—in LIMC social support networks, both locally and transnationally (Fig. 12.2).

Ryan et al. (2008) argue that local and transnational networks offer different types of support. Based on earlier research, these authors distinguish among practical, emotional, and informational support, to which they add companionship and socializing (Ryan et al., 2008). Practical support, such as someone who does your grocery shopping when you are ill, is usually more based on propinquity than emotional support; a listening ear can also be offered by phone or Skype. Socializing, such as

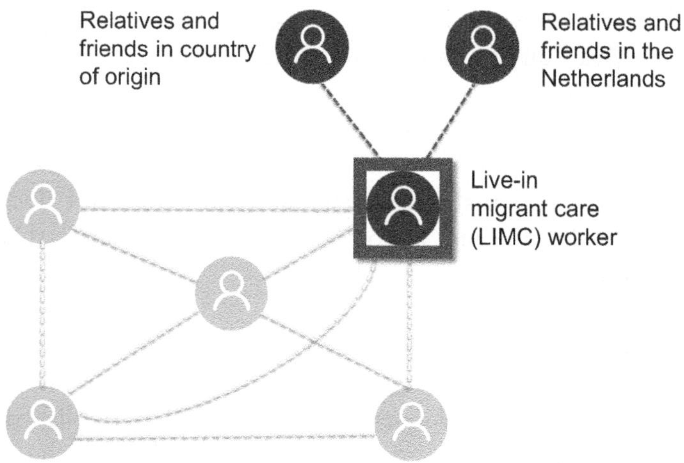

Fig. 12.2 Social support from actors outside the care network

undertaking leisure activities with someone, is also often based on propinquity, while informational support can be derived from both nearby and distant sources. Because local and transnational networks provide different types of support, Boccagni (2014) argues that LIMC workers should be facilitated in accessing and using ICTs for transnational communication, but also in forming local informal networks. Following Van Dijk et al. (2013), we will differentiate between practical support (including informational support) and emotional support (including socializing and companionship).

In sum, we explore the presence or lack of different types of social support that LIMC workers receive 'here' (in the Netherlands) and 'there' (in the home country) from client-related and non-client-related actors.

Live-in Migrant Care Work in the Netherlands

LIMC work is well established in countries with a relatively limited welfare system (e.g., Italy, Spain, and Austria), but is a new development in countries with 'generous' welfare systems, such as Sweden and the Netherlands (Hellgren, 2015; Da Roit & Van Bochove, 2017). Dutch public long-term care (LTC) policies are traditionally among the most inclusive and expensive in the world (Da Roit & Van Bochove, 2017). In recent years, Dutch LTC funding and provision has changed. Access to residential care is now restricted and household help is no longer an individual right (Da Roit & Van Bochove, 2017; Van den Broek et al., 2019). People who need care and support are expected to first mobilize informal networks, before turning to publicly funded services (Van Bochove et al., 2018). Government policies encourage aging in place (Van Dijk et al., 2013). However, informal or professional home care is not always sufficient for people needing almost continuous assistance (Davies & Mans, 2015).

Live-in migrant care (LIMC) intermediary organizations—in other studies also referred to as 'brokering agencies' (Leiber et al., 2021) or 'employment agencies' (Bruquetas-Callejo, 2020)—established in the Netherlands from 2005 onward and promise an alternative to residential care or intensive family care. Compared to 'traditional' home care organizations, these private organizations offer relatively cheap services which clients can generally pay for with publicly funded cash-for-care benefits. Dutch LIMC work is a niche market that has only recently attracted research attention (Van Grafhorst, 2014; Davies & Mans, 2015; Da Roit & Van Bochove, 2017; Horn et al., 2019; Bruquetas-Callejo, 2020). Exploratory research shows that some 15–20 Dutch organizations offer this type of 24-hour care service (Da Roit & Van Bochove, 2017; Bruquetas-Callejo, 2020), employing a few hundred LIMC workers. LIMC workers in the Netherlands are usually women in their 40s, trained nurses, or nursing assistants coming from EU member states in Central and Eastern Europe. The intermediary organizations offer different types of employment modes and rotation turns (Bruquetas-Callejo, 2020). Some LIMC workers are paired in teams, rotating every few weeks or months (one staying with the recipient, the other returning to the home country). Others work alone and stay in the Netherlands for longer periods. Care recipients are usually older people with Alzheimer's or Parkinson's disease. Some are younger adults, suffering from multiple sclerosis for instance (Da Roit & Van Bochove, 2017).

Dutch LIMC intermediary organizations engage in recruitment, matchmaking, training, and supervision. Often, they also offer language courses; most of them do so in the Netherlands but at least one organization does so in the migrants' country of origin (Da Roit & Van Bochove, 2017). Employees of the intermediary organization have regular contact with clients and care workers to check their satisfaction with the working conditions. The research of Da Roit and Van Bochove (2017) suggests that these organizations play an important role in providing practical and emotional support to LIMC workers. How this organizational support relates to other forms of client-related and non-client-related is further explored in this study.

3 Research Methodology

Access to the Field

This study is based on qualitative research conducted among various actors directly involved with Dutch LIMC work, aimed at gaining insight into LIMC workers' wellbeing and precariousness, collaboration in care networks, and transnational social contacts (see Van Bochove et al., 2017; Van Bochove & Zur Kleinsmiede, 2020). As a starting point of the fieldwork, we contacted LIMC intermediary organizations that cooperated in the study of Da Roit and Van Bochove (2017) as well as one additional organization. Because of the relatively small number of LIMC workers in the Netherlands and their lack of visibility in public spaces—combined with their general reluctance to participate in research (Bruquetas-Callejo, 2020)—we depended on the managers of intermediary organizations as gatekeepers to get us in contact with LIMC workers. We asked the managers if we could interview them about the current state of affairs and if they were willing to provide access to other actors involved in care provision, particularly LIMC workers.

Four managers were willing to be interviewed and two of them also provided access to LIMC workers and other actors involved, such as care coordinators working for their organization and informal caregivers. One organization selected six LIMC workers whom we could interview. Through snowball sampling, we selected a seventh one working for the same organization. Another organization asked us to write an email that they could distribute among LIMC workers to ask for their participation, which led to two respondents. The remaining managers said that the participation of LIMC workers was too hard to organize and they needed to protect their workers from public attention.

During one interview with an LIMC worker we also got the opportunity to interview a relative of the care recipients, who also was an informal caregiver. In addition, one district nurse was interviewed. During an interview with an LIMC worker her name and the organization she worked for were mentioned. Afterward she was contacted and she was willing to participate in our research. Interviews with care recipients were not possible, usually because the recipients' medical condition made talking difficult and the managers did not want to bother them. Because of privacy restrictions, conducting observations of clients was not allowed either.

In total, 20 respondents were interviewed: four managers and three care coordinators of LIMC organizations; nine LIMC workers; three relatives of care recipients; and one district nurse. All migrant caregivers were women aged between 26 and 51 who came from Hungary or Romania. Six were paired in teams, rotating every two weeks (one staying with the recipient, the other returning to the home country). The other three did not work in teams and stayed in the Netherlands for longer consecutive periods. In the quotations, in addition to the quoted LIMC worker's age, we indicate whether the respondent was 'paired' or 'non-paired'.

Interviews and Analysis

We interviewed five LIMC workers face to face at the client's home; the other four were interviewed via Skype. Most interviews took place in Dutch, but two were conducted in English, and in one case, an interpreter translated the questions and answers. Interviews with other respondents were conducted face to face or by telephone. Interviews took between 45 minutes and 2 hours.

The interviews were semi-structured, using a topic list with broad topic areas related to wellbeing, precariousness, care networks, and transnational contacts. For the purpose of this chapter, answers to questions like the following were particularly relevant: How do your work activities relate to your leisure time? What do you like about your job and are there aspects of your job that you would like to change or do not like? What kind of places do you visit during/outside work? Can you tell me about the support you receive from friends, family members, the agency, or other workers? Can you describe the relationship you have with your client? Can you describe the relationship with family (or friends) in your home country? How often do you have contact? With whom do you collaborate (e.g., family and nurses)? How do you experience collaborating with formal caregivers and relatives (tensions, positive things, etc.)?

In interviews with respondents other than LIMC workers, we asked them about their contact and/or collaboration with LIMC workers. We allowed respondents to touch upon other themes relevant to them so that new insights could arise (cf. Van Dijk et al., 2013). In this way, we learned about the importance of local contacts outside the care network and the role of different types of employment modes and mobility patterns in shaping social support.

All interviews were tape-recorded and transcribed verbatim. To avoid losing nuance, we did not translate the transcriptions until the reporting stage. Data were analyzed using a preliminary conceptual framework, combining deductive and inductive reasoning (Wilson & Chaddha, 2009). Initial coding was guided by the sensitizing concepts of precariousness, care networks, and transnational ties, focusing on different expressions of precariousness, strategies to deal with precariousness, the composition of care networks, collaboration and tensions between actors in care networks, and contacts with relatives and friends in the home country. These codes were then categorized in secondary themes emerging from the data, such as differences in contract types and related mobility patterns. For the present study, we further zoomed in on the characteristics of social networks in terms of client-related and non-client-related actors, local and transnational contacts, and different types of support (practical and emotional support). We re-read and re-coded the interviews with these themes in mind since this was not the central focus of the earlier analysis.

Ethical Considerations

Before the interviews started, the aim of the study was explained and verbal informed consent was obtained to collaborate in the study and to start recording. We promised anonymity to all our respondents; therefore, we use codes to protect their identities. Particularly when contacting LIMC workers, we emphasized that their participation was voluntary and we assured them that their responses would remain confidential.

4 Findings

Below, we first discuss how actors within care networks can alleviate the working conditions of LIMC workers, by providing practical support (e.g., helping them with care provision) and emotional support (e.g., offering a listening ear). Next, we shift our focus to actors outside care networks, such as LIMC workers' relatives and friends in the country of origin, but also local non-client-related actors. In addition to describing the different types of social support that LIMC workers receive 'here' and 'there' (cf. Boccagni, 2014), we pay attention to factors that influence their access to support.

Social Support Within Care Networks: The Role of Client-Related Actors

Various actors in care networks around recipients of LIMC are potential sources of practical and emotional support for LIMC workers. Our findings suggest that care recipients' relatives often play an important role in care networks and have an enormous impact on how LIMC workers experience their work. In terms of practical support, client relatives are important because they can temporarily substitute the LIMC worker, so she can have some time off. In most of the cases we encountered, LIMC workers received help from family caregivers. For instance, one LIMC worker said that her client's son takes over caring for his father 1.5 days a week. In some cases, however, relatives were not present that often. Some LIMC workers gave examples of children who promised to help, but ultimately did not. One LIMC worker experienced that care recipient relatives caused her more work rather than alleviating her work burden. The client's parents needed care and the LIMC worker often had to look after the client's young child as well.

> The dad wanted a beer every day at 4 pm. I always had to check if his glass was empty and if so, refill it. And I always had to bring the dad to the doctor or hospital. (LIMC worker 6, aged 39, paired)

This LIMC worker complained about the situation to the intermediary organization, which arranged another client for her.

Some LIMC workers said they had family-like relationships with clients' relatives, referring to a care recipients' son as 'brother' or to the care recipient's wife as 'grandmother'. One of these LIMC workers said: 'I am lucky. In the morning, I get a morning kiss and in the evening, a good-night kiss. It is just like being with my grandparents' (LIMC worker 7, aged 26, non-paired).

While Salami et al. (2017: 1676) mention the risks of being treated 'like one of the family'—it makes it easier to exploit LIMC workers—the workers we interviewed usually found family-like bonds soothing rather than burdensome. In such cases, LIMC workers also received emotional support from care recipients' relatives. One of the relatives we interviewed said that he and his family urged an LIMC worker to go back to her home country when her father had a cerebral hemorrhage. Although the LIMC worker felt bad about it, they told her to 'make the right decision for your own family, not ours' (care recipient's relative 1). In this example, the client's relatives took care of the client 'here', to enable the LIMC worker to care for someone in need 'there'. Client relatives may also contribute to the local incorporation of LIMC workers, for instance by inviting them for birthday parties and going on day trips. The LIMC worker quoted earlier about her negative experience with client relatives also had very positive experiences. She talked about her bond with the daughter of another former client, who provided both practical and emotional support.

> She always asked me if she could help and what she could do. And we always undertook activities. We went to Keukenhof [the large flower garden in the Netherlands] and to the zoo many times. I really enjoyed that. It was the best. And now [after the client passed away and the respondent works for a new client] she is my friend, a real good friend. (LIMC worker 6, aged 39, paired)

As this example shows, the distinction between actors within and outside care networks may be blurred: a client-related actor can still provide support to the LIMC worker after the client-caregiver relationship ended.

Although less common, some LIMC workers also became friends with other paid caregivers. For instance, one LIMC worker said a district nurse, who also visits the respondent's client, has become her friend: 'I was at her place this Sunday, she invited me' (LIMC worker 7, 26, non-paired). LIMC workers who are paired in teams, rotating every two weeks, see each other on the 'changeover day' and often also have contact in between, by phone or email. They discuss relevant practical issues related to the client, but also their personal circumstances. Two paired

LIMC workers (LIMC workers 3 and 4) who were interviewed together said that they call each other when they find the work burdensome, for instance when they have a headache. They pointed out that this mutual emotional support is very important to them. For non-paired LIMC workers, contact with co-workers is more difficult to arrange. Often, they do know co-workers in the Netherlands—for instance, because they stayed in touch after following a Dutch language course in the home country, organized by the intermediary organization—but as they work in different parts of the country and have only limited time off, meetings are difficult to arrange.

In all four LIMC intermediary organizations involved in this research, employees of the organization, usually care coordinators, have regular evaluation meetings with care recipients (and their relatives) and LIMC workers, to check if everyone is still satisfied. One of the managers (Manager 2), who regularly talked with LIMC workers himself, gave an example of an LIMC worker who, when asked how things were going, first said everything was alright. But she had tears in her eyes. The manager then asked a care coordinator to talk to her and it turned out that the LIMC worker missed her boyfriend and her relatives 'back home'. She worked in the Netherlands for longer consecutive periods, but she rather wanted to work in pairs and rotate every two weeks. The organization then changed her contract and teamed her up with another LIMC worker. This example shows that practical support 'here' can also enable emotional support 'there'. It also indicates the importance of care coordinators in LIMC organizations. Various managers said that LIMC workers are less likely to share their personal stories with their 'boss' (i.e., the manager) because they do not want to be seen as difficult.

In sum, the findings show that particularly care recipients' relatives can be important sources of practical and emotional support. When relatives are largely absent, or present in a negative sense, LIMC workers are more likely to experience burdensome situations. Differences between LIMC workers regarding their client-related contacts did not appear to be related to the type of contract they have—rotating or staying for longer periods. Organizational policies and practices are important in a different way, however. LIMC intermediary organizations can play a mediating role in solving unfavorable situations. Moreover, it is the organization's responsibility to screen potential clients and their care networks on the suitability for receiving LIMC and assess current working conditions on a regular basis.

Social Support Outside Care Networks: Relatives and Friends

Not all LIMC workers have bonds with other actors in the delivery of care around the care recipient. Some LIMC workers feel isolated in the home of the client or feel overburdened by the demands of client relatives. Particularly in such cases, a non-client-related social network is important to be able to cope with the job.

All LIMC workers said they contact relatives and friends in the country of origin almost daily by phone and via Skype or other messaging systems. They find it very important to stay in touch; some of them contact their family every morning and evening, before and after work. As LIMC worker 5 (aged 43, paired) put it, it is important for her 'soul'. Another LIMC worker said:

> I phone my husband every day. I tell him 'I am a bit tired', 'I am happy now', 'I have pain' or 'I am well'. I also ask him for advice in private matters, and so on. (…) I don't have family here, I am alone, so it is very important to have people to talk to. (LIMC worker 3, aged 26, paired)

The experience of another LIMC worker (6, aged 39, paired) shows how problems within the care network can also affect access to support outside the care network. During her stay with the client who had demanding parents—the dad who wanted her to serve his beer every day—it was difficult for her to stay in touch with her relatives. Not only because the internet connection was poor, but also because the parents found it problematic that she talked to them in her own language. This made the already bad experience there even worse.

How often LIMC workers visit their country of origin is related to the type of contract they have. Most of the LIMC workers we interviewed (six out of nine) were paired in teams and worked in shifts of 'two weeks on, two weeks off', as Manager 2 called it. During the two weeks that they do not have to work, these LIMC workers return to their country of origin. The other three respondents stayed in the Netherlands for longer periods and visited their home country during holidays and sometimes during weekends.

In line with the findings of Bruquetas-Callejo (2020), LIMC workers who rotate describe this type of contract as a strategic choice: two weeks of hard work in the Netherlands are compensated by two weeks of quality time with their relatives and friends back home. Managers of intermediary organizations that offer both types of contracts also emphasize the advantages for the wellbeing of LIMC workers (cf. Da Roit & Van Bochove, 2017). According to them, LIMC workers can use the weeks in their home country to 'empty their heads'; they 'return [to the Netherlands] completely refreshed'; and it has 'tremendous advantages' for them to return regularly to their trusted environment where they can speak their mother tongue.

Despite the possible benefits for LIMC workers, this type of contract also has its drawbacks. LIMC workers that were paired in teams often mentioned that they experienced their stay in the Netherlands as burdensome because they do not have a real private life here:

> My life is not here. I only work here; my personal life is not here. (...) I don't know the Netherlands. For example, I can visit family with the client but I can't visit [city name] on Friday or Sunday by myself, that is impossible. It feels like I am locked here. (...) Everyone tells us [two paired LIMC workers who were interviewed together] how great it is: two weeks of work, and then two weeks off. But it is not great, it actually is a bit difficult. The difficult thing is that I am here for two weeks, but I do not have a day off. I can go grocery shopping, but that still is work. (LIMC worker 3, aged 26, paired)

During their stay with the client, they are focused on their work and, next to some client-related social contacts, mainly turn to relatives back home for support. While rotating may be beneficial for sustaining existing ties in the country of origin, it may hinder building up new social relationships, not only 'here' but also 'there'. LIMC workers who had a partner often mentioned how important this was for them because they could share their experiences and talk about problems. Their partner helped them to cope with the more burdensome aspects of their work. The LIMC worker quoted below does not have a partner and thinks that rotating between the Netherlands and her home country hinders finding one:

> I am two weeks here and then two weeks in Budapest. Because of that, I cannot build a relationship with someone. That is a bit of a problem. I want to have children and a relationship, but because I am two weeks here, and two weeks there, that is a bit difficult. (LIMC worker 1, age unknown, paired)

LIMC workers 7 and 8, both non-paired workers, have the opportunity to spend time with their relatives during their stay in the Netherlands. LIMC worker 7 is the daughter of LIMC worker 8. Their clients live about 25 km apart and they are able to undertake activities together when they have time off. The daughter said that also during work time, she talks with her mother as often as she wants. Being trained as a medical nurse, she sometimes gives her mother practical advice. Her mother in turn gives her moral support. Her partner, whom she met in the country of origin, works as a truck driver and regularly visits the Netherlands for work. During these visits, they are able to spend time together.

> And the family loves him [LIMC worker's partner] as well. (…) When he came here he surprised me. He booked a hotel room and the family asked: 'Why did he book that hotel when we have plenty of space here?' (…) And the lady loves him, she is kissing him also. (LIMC worker 7, aged 26, non-paired)

The three non-paired workers who participated in this research also had more time to invest in local contacts. LIMC worker 8 said that she feels a strong affinity with the Netherlands and she plans to stay here, also because she has a Dutch partner:

> I can stay as long as necessary because I like the Netherlands. I have a very good relationship for more than eight months with my boyfriend and I'm really planning to stay because it's totally different, it feels like home. (LIMC worker 8, aged 51, non-paired)

LIMC worker 9, who has been in the Netherlands for three years and currently lives in with a client in Amsterdam, says she has a 'wide circle of acquaintances' and regularly goes to the movies or to a museum with them (LIMC worker 9, aged 27, non-paired). When she faces a work-related problem or dilemma, she not only asks her family 'back home' for their opinion, but also some of her close friends in the Netherlands. Having a private life in the Netherlands, knowing other 'locals', and participating in leisure activities with them helped non-paired workers to cope with the emotional and physical demands of their job. Compared with paired workers, who—similar to LIMC worker 3 (aged 26, paired)—often experienced to be 'locked in' during their stay in the Netherlands, non-paired workers felt more at ease and at 'home' there.

In sum, the findings suggest that differences between LIMC workers' access to non-client-related social support are related to the type of contract they have. LIMC workers who work in pairs return to their home country two weeks per month and are able to spend much time with their loved ones there. However, during their stay in the Netherlands, they do not have much time off and are not able to build up a local non-client-related network. Such a network often provides access to practical and emotional support and mitigates the experience of precariousness. The non-paired LIMC workers included in this research do have contacts with actors outside the care network, which helps them to feel comfortable in the Netherlands.

5 Discussion and Conclusion

The findings of this study suggest that both client-related and non-client-related social support networks play an important role in shaping LIMC workers' working conditions and in how they experience their job. Some LIMC workers are part of well-functioning care networks, in which care recipients' relatives and formal caregivers substitute LIMC workers so they have time off. However, other LIMC workers do not receive practical and emotional support from client relatives, or worse, client relatives enhance the experience of precariousness by making unreasonable demands. For all LIMC workers, but particularly for those who do not receive support from client-related actors, a non-client-related social network is vital to be able to cope with emotionally challenging aspects of their work. While all LIMC workers have regular contacts with relatives and friends in their country of origin, only LIMC workers who work in the Netherlands for longer consecutive periods have sufficient time to build up a local non-client-related support network. LIMC workers who rotate every two weeks are able to visit their home country regularly and can spend more time with their relatives and friends 'there', but more often feel isolated 'here', all the more so when they do not feel supported by client-related actors. Similar to the findings of Wee et al. (2019), our findings suggest that various interrelated conditions shape the experience of a higher or lower degree of precarious work.

Two results of our exploratory study are worth further investigation. First, while in existing literature on care networks a strict boundary is drawn between actors who provide care for a care recipient on a regular basis and actors outside the care network, our findings suggest that care networks and broader social support networks are often intertwined in various ways. By providing practical and emotional support to LIMC workers, actors outside care networks help sustain the practice of caregiving within care networks. The boundaries between client-related and non-client-related networks are sometimes also blurred because within care networks, friendships can develop that last longer than the original care-related relationship. While a focus on care networks rather than on dyad (or triad) relationships between LIMC workers and care recipients (and their relatives) is already a step in the right direction (Van Bochove & Zur Kleinsmiede, 2020), to do justice to the complex entanglement and involvement of a growing number of actors in LIMC work, it is important to also include LIMC workers' broader social support networks.

A second insight yielded by our research is that LIMC intermediary organizations can play an important role in shaping the conditions of LIMC work. Within care networks, the intermediary organizations that participated in our research in the Netherlands try to make a good match between LIMC workers and clients and their relatives and regularly evaluate the care practice to see if everyone is still satisfied. When intermediary organizations do not identify potential risks in time, this may lead to precarious situations (for similar findings on Germany, see Leiber et al., 2019). Moreover, and perhaps more surprisingly, we found that intermediary organizations also influence LIMC workers' access to support outside care networks. The different employment modes that various organizations offer (paired and non-paired work) largely shape LIMC workers' mobility patterns and consequently their access to local and transnational support. While LIMC workers make the decision for one type or the other themselves—many of them strategically choose for working in pairs, to be able to spend more time in the country of origin—LIMC organizations are responsible for how these different contract types work out in practice.

LIMC workers who stay in the Netherlands for longer periods are not necessarily better off in terms of job demands and job resources than those who travel back and forth. How they experience their working conditions depends on various factors, including the pursuit of 'personal life-projects' in the country of origin (Bruquetas-Callejo, 2020) and the support they receive from actors in the care network. However, no matter how warm and supportive, close informal relations between LIMC workers and care recipients and their relatives are still based on asymmetrical power relationships between 'worker' and 'employer' (Salami et al., 2017). In dealing with physically or emotionally harsh working conditions, it seems important that LIMC workers have a non-client-related support network in the country of work. Rotating every two weeks and having almost no time off during workweeks makes it extremely difficult to build such a local support network. Managers of intermediary organizations who offer paired work emphasized the advantages for LIMC workers. However, it is important that managers acknowledge that it hampers local embeddedness of LIMC workers. Even though many LIMC workers do not plan to stay in the Netherlands permanently, they often still wish to get to know the country and the people better. Guaranteeing more time off during the workweeks would be a first step in this direction.

Our study has some limitations. First, we did not speak to all relevant actors within and outside the care networks of which LIMC workers are part. In some cases, we talked with formal caregivers, in other cases with informal caregivers, and in yet other cases, we could only interview LIMC workers. We did not interview LIMC workers' relatives or friends.

Second, in some interviews, the language barrier made it difficult to discuss personal issues such as emotions in depth. An interpreter assisted in one interview, but this negatively affected the spontaneity of the interviewee. However, most LIMC workers spoke Dutch relatively well and two were fluent in English. Third, the LIMC workers we interviewed worked for two intermediary organizations, while there are some 15–20 organizations in the Netherlands offering this type of service, each with their own characteristics. We focused on differences between LIMC workers' local and transnational networks based on the type of contract they had. If we had been able to include LIMC workers from other organizations as well, we might have found more variety in the types of bonds and composition of networks. Finally, by focusing on how support from actors within and outside LIMC workers' care networks may mitigate their experience of precariousness, the emotional and practical support that LIMC workers provide themselves in these networks largely remained outside the scope of this study.

The interlinkages between care networks and migrants' social support networks deserve further attention in future research. In addition, it would be interesting to study how the characteristics of intermediary organizations—for instance, their size and risk management strategies (Leiber et al., 2019)—affect the local and transnational incorporation of LIMC workers.

References

Bauer, G., & Österle, A. (2013). Migrant care labour: The commodification and redistribution of care and emotional work. *Social Policy and Society, 12*(3), 461.

Boccagni, P. (2014). Caring about migrant care workers: From private obligations to transnational social welfare? *Critical Social Policy, 34*(2), 221–240.

Broese van Groenou, M., Jacobs, M., Zwart-Olde, I., & Deeg, D. J. (2016). Mixed care networks of community-dwelling older adults with physical health impairments in the Netherlands. *Health and Social Care in the Community, 24*(1), 95–104.

Bruquetas-Callejo, M. (2020). Long-term care crisis in the Netherlands and migration of live-in care workers: Transnational trajectories, coping strategies and motivation mixes. *International Migration, 58*(1), 105–118.
Da Roit, B., & van Bochove, M. (2017). Migrant care work going Dutch? The emergence of a live-in migrant care market and the restructuring of the Dutch long-term care system. *Social Policy & Administration, 51*(1), 76–94.
Da Roit, B., & Weicht, B. (2013). Migrant care work and care, migration and employment regimes: A fuzzy-set analysis. *Journal of European Social Policy, 23*(5), 469–486.
Davies, A., & Mans, L. (2015). Thuisbezorgd—Inwonende buitenlandse zorgverleners als oplossing? *Podium voor Bio-ethiek, 22*(2), 23–26.
Finch, B. K., & Vega, W. A. (2003). Acculturation stress, social support, and self-rated health among Latinos in California. *Journal of Immigrant Health, 5*(3), 109–117.
Hellgren, Z. (2015). Markets, regimes, and the role of stakeholders: Explaining precariousness of migrant domestic/care workers in different institutional frameworks. *Social Politics: International Studies in Gender, State and Society, 22*(2), 220–241.
Horn, V., Schweppe, C., Böcker, A., et al. (2019). Live-in migrant care worker arrangements in Germany and the Netherlands: Motivations and justifications in family decision-making. *International Journal of Ageing and Later Life, 13*(2), 83–113.
Jacobs, M. T., Broese van Groenou, M. I., Aartsen, M. J., & Deeg, D. J. (2016). Diversity in older adults' care networks: The added value of individual beliefs and social network proximity. *The Journals of Gerontology: Series B, 73*(2), 326–336.
Kalleberg, A. L. (2009). Precarious work, insecure workers: Employment relations in transition. *American Sociological Review, 74*(1), 1–22.
Kinman, G., & Leggetter, S. (2016). Emotional labour and wellbeing: What protects nurses? *Healthcare, 4*(4), 89.
Leiber, S., Matuszczyk, K., & Rossow, V. (2019). Private labor market intermediaries in the Europeanized live-in care market between Germany and Poland: A typology. *Zeitschrift für Sozialreform, 65*(3), 365–392.
Leiber, S., Rossow, V., Österle, A., & Frerk, T. (2021). Yet another black box: brokering agencies in the evolving market for live-in migrant care work in Austria and Germany. *International Journal of Care and Caring, 5*(2), 187–208.

Lutz, H. (2016). When home becomes a workplace: Domestic work as an ordinary job in Germany? In H. Lutz (Ed.), *Migration and domestic work* (pp. 55–72). Routledge.

Oxman-Martinez, J., Hanley, J., & Cheung, L. (2004). *Another look at the live-in-caregivers program: An analysis of an action research survey conducted by PINAY, the Quebec Filipino women's association with the centre for applied family studies*. Centre de recherche interuniversitaire de Montréal sur l'immigration, l'intégration et la dynamique urbaine.

Salami, B., Duggleby, W., & Rajani, F. (2017). The perspective of employers/families and care recipients of migrant live-in caregivers: A scoping review. *Health and Social Care in the Community, 25*(6), 1667–1678.

Ryan, L., Sales, R., Tilki, M., & Siara, B. (2008). Social networks, social support and social capital: The experiences of recent Polish migrants in London. *Sociology, 42*(4), 672–690.

Tonkens, E. (2012). Working with Arlie Hochschild: Connecting feelings to social change. *Social Politics, 19*(2), 194–218. https://doi.org/10.1093/sp/jxs003

Van Bochove, M. (2012). *Geographies of belonging: The transnational and local involvement of economically successful migrants*. Dissertation. Erasmus University, Rotterdam.

Van Bochove, M., & Zur Kleinsmiede, D. (2020). Broadening the scope of live-in migrant care research: How care networks shape the experience of precarious work. *Health and Social Care in the Community, 28*(1), 51–59.

Van Bochove, M., Zur Kleinsmiede, D., & Ashu, S. (2017). *Kwetsbaar, en dan? Hoe inwonende buitenlandse zorgverleners en hun werkgevers omgaan met precaire arbeid*. Erasmus University Rotterdam.

Van Bochove, M., Tonkens, E., Verplanke, L., & Roggeveen, S. (2018). Reconstructing the professional domain: Boundary work of professionals and volunteers in the context of social service reform. *Current Sociology, 66*(3), 392–411.

Van den Broek, T., Dykstra, P. A., & Van Der Veen, R. J. (2019). Adult children stepping in? Long-term care reforms and trends in children's provision of household support to impaired parents in the Netherlands. *Ageing & Society, 39*(1), 112–137.

Van Dijk, H. M., Cramm, J. M., & Nieboer, A. P. (2013). The experiences of neighbour, volunteer and professional support-givers in supporting community dwelling older people. *Health and Social Care in the Community, 21*(2), 150–158.

Van Grafhorst, A. (2014). *Verantwoorde buitenlandse zorg aan huis: Een verkennend onderzoek naar inwonende buitenlandse zorgverleners in Nederland.* Amsterdam, Netherlands: Stichting Wemos.

Wee, K., Goh, C., & Yeoh, B. S. (2019). Chutes-and-ladders: The migration industry, conditionality, and the production of precarity among migrant domestic workers in Singapore. *Journal of Ethnic and Migration Studies, 45*(14), 2672–2688.

Wilson, W. J., & Chaddha, A. (2009). The role of theory in ethnographic research. *Ethnography, 10*(4), 549–564.

13

Elder Abuse in Live-in Migrant Carer Arrangements

Vincent Horn and Cornelia Schweppe

1 Introduction

Internationally research on elder abuse in care at home and research on live-in migrant carers (LIMCs) have developed widely during the last decades. So far, however, these two research strands have not been connected to each other. Research on elder abuse in home care settings does not examine LIMC arrangements and research on LIMC arrangements does not turn to the question of elder abuse. One of the few exceptions is the qualitative study by Liat Ayalon (2009, 2011) on abuse and neglect in private Israeli households employing an LIMC. Considering the high potential for abuse in old age care at home, the lack of attention paid to

This chapter is a modified version of the article: Horn, V., & Schweppe, C. (2019). Gewaltpotentiale und Gewaltschutz in Privathaushalten mit 24-Stunden-Pflegekräften. *Zeitschrift für Sozialpädagogik, 18*(4), 394–413.

V. Horn (✉) • C. Schweppe
Institute of Education, Johannes Gutenberg University of Mainz, Mainz, Germany
e-mail: hornv@uni-mainz.de; c.schweppe@uni-mainz.de

V. Horn et al. (eds.), *The Global Old Age Care Industry*,
https://doi.org/10.1007/978-981-16-2237-3_13

this subject in LIMC settings comes as a surprise. Internationally, elder abuse in old age home care has become a wide field of research with a special focus on the prevalence and risk factors of elder abuse in family old age care relationships (e.g. Cooney et al., 2006; Orfila et al., 2018; Wang et al., 2018). Many studies indicate that abuse against elder care recipients is a widespread phenomenon in old age home care (e.g. Cooper et al., 2009; Eggert et al., 2018; Görgen et al., 2012; Graß et al., 2007; Wiglesworth et al., 2010).

This chapter addresses this gap by looking into the factors, which might increase the risks of elder abuse or protect against it in LIMC arrangements in Germany. These care arrangements, in which older people in need of care are supported, accompanied and cared for by migrant carers mainly from Central and Eastern European countries, have become a significant pillar of the old age care system in Germany (Arend & Klie, 2017; Neuhaus et al., 2009).

The empirical findings presented in this chapter are based on the research project "Emergence and significance of transnational care arrangements" (ESTRANCA), which was carried out in cooperation with the University of Mainz (Germany) and Radboud University (The Netherlands).[1] Against the background of the different old age care regimes in Germany and the Netherlands, the aim of ESTRANCA was to gain a deeper understanding of the development and significance of LIMC arrangements in both countries (see Horn et al. in this book). The research project did not focus on issues around elder abuse and protecting against it. However, during our analysis, related aspects repeatedly emerged. We analyzed them in further detail in the course of a secondary analysis based on the data we collected in Germany. The findings are presented in this chapter.

2 24-Hour Caregivers in the German System

The emergence of LIMC arrangements in Germany can be explained against the background of the pronounced priority given to old age care at home over residential care. Home care in Germany is considered as "good" and, in relation to residential old age care, as "better." It places a high responsibility on family members and demands a high level of their involvement in old age care. The priority of old age care at home over residential care is anchored in German social law. In addition to idealizing concepts of the family and moralizing discourse about the family's responsibilities to care for their elder relatives, high private costs and the poor quality of care provided in residential care play a role in the fact that, according to opinion polls, the preference for home care is deeply rooted in Germans' attitudes (see Böcker et al., 2017 and Horn et al. in this book).

At the same time, however, old age home care is increasingly faced with considerable challenges due to changes in the family structures (e.g. increase in women's employment, the absence of or geographical distance from family members), the increasing length of time that old age care is required and the changing care requirements, for example, due to dementia. A large number of studies reveal the high psychological, social, physical and financial burdens placed on family carers (e.g. see Bestmann et al., 2014; Pinquart, 2016). Outpatient old age care services in Germany can mitigate those burdens only to a limited extent. Due to the financing modes of the long-term care insurance in Germany, they can often only be used selectively for nursing-related or household-related support.

In this conflicting situation of placing a high responsibility of old age care on the family while at the same time overburdening the family, the employment of an LIMC presents a suitable solution. The following quote from an article on the German news portal NRZ summarizes the situation well: "When no-one knows what else to do, someone is bound to comment, 'Then get a Pole in!' Poles are the nation's most important stopgap" (Maibaum, 2017). The fact that LIMCs are recruited from abroad is easily explained: there are no affordable alternatives in Germany. The LIMCs' cross-border mobility is encouraged by the freedom of movement for workers in the EU, successively extended to the new acceding countries during its eastern enlargement of 2004.

The extent of old age home care in Germany is reflected by the care statistics compiled by the German Federal Statistical Office (Statistisches Bundesamt, 2017). However, it is important to note that these statistics only cover those persons in need of care receiving benefits from the German long-term care insurance. According to the statistics, 2.1 out of the roughly 2.9 million people in need of care in Germany are cared for at home. Most of them (1.64 million) are aged 65 or older and only receive financial benefits from long-term care insurance. In other words, in many private German households, older people are cared for by family members alone without the support of outpatient old age care services. In 2017, a representative survey commissioned by the Hans Böckler Foundation found that 11% of these private households had employed an LIMC (see Hielscher et al., 2017). This would mean that there are currently around 180,000 private households in Germany that rely on an LIMC.

However, the actual number of LIMCs is likely to be significantly higher, firstly because LIMCs often work in rotation, replacing one another at intervals of one to three months. The number of LIMCs is thus significantly higher than the number of private households that choose LIMC arrangements. Secondly, the number of LIMCs is likely to be higher because of their often irregular mode of employment (Kniejska, 2016). In such cases, the willingness to provide information about this type of care arrangement is likely to be limited, even in anonymized surveys.

A second option to employ an LIMC is via one of the placement agencies which have been established widely in Germany. According to an inquiry by the German consumer organization Stiftung Warentest (2017), around 250 placement agencies are currently working in the field of 24-hour old age migrant care arrangement in Germany. The umbrella organization for several of the larger placement agencies estimates that roughly 30,000 private households currently use the services of an agency (see Horn et al., 2019). Agencies frequently cooperate with companies in other countries, where the LIMCs are officially employed (see Krawietz, 2014). One widespread model is for LIMCs to be posted to private German households for periods of up to three months via the EU Posting of Workers Directive (see Rossow & Leiber, 2017 and Leiber and Rossow in this book).

The cost of an LIMC employed via an agency varies widely but is generally much higher than for a carer who is employed irregularly. The costing is based on various factors, but the level of proficiency in the German language is a key factor. Employment through a placement agency, which is associated with higher costs, does not result in better wages for the

LIMCs. The difference instead goes to the agencies, which usually generate their profits through monthly fees. Family members wanting to avoid this practice have to take on the role of an employer themselves by hiring the LIMC as a domestic help. In this case, they have to pay income tax and social security contributions in line with German legislation. However, the associated costs and the bureaucracy required to register the domestic help make this form of employment less popular.

LIMC arrangements in Germany have been analyzed from different angles. In addition to studies that analyze the different employment models or the practices of the placement agencies (see Böning & Steffen, 2014; Krawietz, 2014; Rossow & Leiber, 2017; Tießler-Marenda, 2012), there are studies on LIMCs' experiences, transnational relationships and working conditions (Apitzsch & Schmidbauer, 2010; Ignatzi, 2014; Karakayali, 2010; Kniejska, 2016; Metz-Göckel et al., 2006; Satola, 2015; Scheiwe & Krawietz, 2010). In terms of working conditions, particular attention has been given to the often-precarious employment conditions. While German labor law provides for a maximum of 48 hours a week, including a day off, a non-representative study of around 900 LIMCs shows that a quarter works 11 or more hours per day (cf. Petermann et al., 2017). Seventy percent stated that they had no day off during the week.

3 Abuse in Old Age Care

Following the World Health Organization's definition of elder abuse, Konopik and Oswald (cited in Konopik et al., 2018: 21) understand abuse in old age care as "a single, or repeated act, or a lack of appropriate action, occurring within a care relationship which causes harm or distress to an older person or to the caring person." This definition is based on a concept of abuse that includes forms of mistreatment as well as neglect. Abuse can arise from either performing or failing to perform an action and can be intentional or unintentional. Abuse includes physical (e.g. hitting someone), psychological (e.g. insults or intimidations) as well as care related (e.g. not providing help, misjudgment of care needs) and emotional or psychosocial neglect (Dieck, 1987; Görgen, 2017). Abuse also involves avoidable measures that restrict a person's autonomy with respect to their freedom to move, freedom of action and freedom of decision-making, including steps taken to deprive them of their liberty or exploit them financially. Abuse in care can affect both the recipient and the caregiver (Pillemer & Suitor, 1992).

The conditions under which abuse emerges in old age care are complex. The following main risk factors for elder abuse in home care settings have been identified, which can occur in varying combinations (e.g. Buchegger-Traxler, 2016; Hirsch, 2011; Kühnert, 1997; Lachs & Pillemer, 2015; Müller-Hergl, 2011; Pillemer & Suitor, 1992; Pillemer et al., 2016; Rabold & Görgen, 2007; Weissenberger-Leduc & Weiberg, 2011; WHO, 2002): They particularly focus on elder abuse that emerges in family care relationships and includes:

- Caregiver's stress and overburdens, for example, due to a high level of care and support needs, or the permanent dependency of the person in need of care;
- Lack of caregiving skills, for example, due to a lack of knowledge about the symptoms and stages of a disease;
- A problematic relationship between the care recipient and the carer before entering into a care relationship, for example, due to unresolved family or partnership conflicts or problematic parent-child relationships;
- Alcohol and substance abuse, for example, as an attempt to deal with stressful situations;
- Dementia: older people suffering from dementia or other mental illnesses are considered to be particularly at risk;
- Abuse toward the caregiver by the care recipient, which in turn can cause the carer to abuse the care receiver.

4 Methodological Design

Our analysis is based on semi-structured interviews with 14 family members in Germany who were involved in care and/or in organizing the LIMC arrangements. They were conducted between 2016 and 2019 at various locations. The interviews focused on the reasons why an LIMC was employed and the everyday life of these care arrangements, including the organization of the care arrangements and the nature of the relationships which emerged in these settings. Access to the interviewees was mainly gained via personal contacts and various gatekeepers (e.g. employees/staff members of old age care services). If possible, interviews were also carried out with the care receivers and/or the LIMC. In addition, five telephone interviews were conducted with the managers of various placement agencies, focusing on questions how migrant care workers were recruited, placement services and forms of employment. Finally, four focus groups were carried out with professionals of outpatient old age care services, each with three to seven participants. The focus groups centered on the professionals' experiences in households that employed an LIMC. All the interviews were recorded with audio devices and afterward transcribed.

The data analysis followed a two-stage approach. First, the factors that emerged in the first analysis concerning risks of and protection from abuse were systematically applied to the data in the secondary analysis and analyzed in that regard. Second, to sharpen the focus further, the data were also analyzed from the point of view of the central risk factors for abuse in home care, as identified in the research (see above). The relevant text passages from the different interviews (with family members, agencies, etc.) were sorted into a provisional category system. This was subsequently consolidated by merging closely related categories.

5 Empirical Findings

Overburden and Relief

As caregiver's stress and overburdens have been identified as a risk factor of elder abuse, in the following we will analyze insofar as LIMC arrangements might lower these burdens and therefore may contribute to reducing the risk of elder abuse.

Being overburdened was one of the main reasons why family members decided to employ an LIMC (see Horn et al., 2019). The question if the stress and burden placed on the family members are reduced in LIMC arrangements and the care recipients is complex. This may certainly be the case. However, our data show that employing an LIMC by no means always and directly result in relief for the family members. Our data revealed that family members often remain closely involved in the care of the older family member or that home care would not have been possible without their participation.

A high level of involvement particularly concerned family members of relatives who were in high need of care. In these cases, a single worker was often not sufficient. Due to the intensive care involvement, the carers' daily routine was structured and timed around their relatives' need for care, despite the employment of an LIMC. Longer absences, for example, holidays, and even shorter absences during the day, had to be planned precisely and could only be realized if the family members could find a replacement. This can be illustrated by the example of Mr. Haupert, aged 70. Due to the very high level of care needed for his wife, the constant presence of two carers was required. It was delivered by him together with the LIMC. He explained that the "bad thing" was that his wife's care constantly required two people: "There always have to be two of us, one person just can't do it anymore." Accordingly, for him it is not the work that is so "bad," but the need to be present at all times: "It's not the work, that's not even that bad, it's just that you always have to be there (emphasized)." The permanent presence of two people required continuous "organizing" and the constant need of having to find replacements in the event of a sudden unavailability of one of the LIMCs.

The data also showed that when an LIMC was employed, family members were confronted with new tasks which were often time-consuming and stressful. This included the organization and supervision of the LIMC arrangement. In addition to administrative tasks such as contractual matters or applying to receive benefits from the long-term care insurance, the need to familiarize the LIMC with the household and the care recipients' needs and desires and dealing with and resolving conflicts between the LIMC and the care recipients were experienced as especially time-consuming and stressful.

The relation between stress and overburden and the risk of abuse is by no means restricted to care burdens of the family members. The often-precarious employment conditions and high workload described above highlight points to the (potential) overburdens of the LIMC. As an example of what the work of an LIMC can imply, Ms. Lingert talks about the working hours of the LIMC who takes care of her mother: "There are no mornings and there are evenings, there are simply 24 hours of the day." Another aspect, which must be taken into account, is the stress-generating factors that may arise from the specifics of the care arrangements, such as social isolation, recurring routines and the feeling of being locked up in a single place. One LIMC compared her situation before leaving for Poland with that of a nun in a convent: "Do you know what I mean? Now the nun is leaving the convent. You can laugh about it, but it's true? We stay here all day (…), don't go to the cinema … there's no man around (laughs). It's like nuns in a convent." Many family members, both children and spouses, often recognized how challenging the LIMC's work was and look for ways of relieving them and thus try to reduce stress and overburdens.

One way in which relatives tried to relieve the LIMC was by taking over some of the care tasks themselves and/or by hiring in additional care services. One family member explains the outpatient care service that comes three times a week: "I have to give them [LIMCs; the authors] time off too; they can't work round the clock all day, that's impossible." When the care arrangement worked well from the relatives' point of view, they were often also willing to respond to the demands of the LIMCs. Several family members reported that they had given in to demands to raise their wages in order not to jeopardize the care arrangement. Others voluntarily paid more money or offered gifts to secure the services of LIMCs, or to show their gratitude.

Even if such practices can contribute to relief and alleviation, there are also clear limits as they often have little effect on the constitutive structure of the LIMC's working conditions (e.g. irregular employment, insufficient separation between work and time off). Moreover, practices of this kind usually come at an additional cost and thus depend (among other things) on the family's financial resources. Our data show that due to cost reasons, some families cut down on or refrain entirely from using additional professional care services, in part as a means of being able to afford the 24-hour caregiver. In view of the high costs, hiring a second 24-hour caregiver to relieve the first one at night—a practice we came across in one of our cases—would be an exception, limited to families with relatively high financial resources.

The LIMCs do not rely solely on the initiative of relatives when it comes to looking for means of relieving the burden and improving their working conditions: they also take action themselves. The data show that the 24-hour caregivers are generally very much aware of their market value and how important they are to the family. This awareness opens up new avenues of action for them. This can be seen firstly from the fact that they are not willing to accept any job that is offered to them. The managing director of an agency reported that it was becoming increasingly difficult to find LIMCs to work in rural and remote areas. It can also be seen from the fact that LIMCs ask their employers for pay increases and leave care arrangements which are not acceptable to them. This indicates that, due to the high demand, they can leave one position without having to wait long for another.

Lack of Qualifications

Most of the LIMCs did not have any training in care for the sick or elderly, but instead relied upon the experience they had gained from looking after family members and/or working in Germany. The placement agencies also did not require them to have trained or studied in the healthcare sector. As the managing director of one placement agency said, "That's why a lot of it is empathy; so it might be a teacher who makes the perfect carer, we might have someone who may have been a cook or something; there are no requirements when it comes to the training." Interestingly, the family members also placed less emphasis on professional characteristics and more on personality traits. In response to the question of what was important to relatives when selecting an LIMC and assessing the care arrangement, the answer "The chemistry has to be right" was given repeatedly in the interviews.

Some studies have found that insufficient knowledge about nursing or illnesses can lead to (unintended) mistakes and to carers being overburdened (Nägele et al., 2010). This was also confirmed in our study, especially in the interviews with professionals working for outpatient old age care services, who encountered serious pressure ulcers caused by malpositioning on a relatively frequent basis in households employing an LIMC. At the same time, a lack of qualifications can have serious psychological implications for the LIMCs:

And I think there's a difference when you just look after someone a bit, help out in the household and go shopping, or when you really have someone in extreme need of care lying there. That was the case with them; sometimes they were totally out of their depth. They were in tears because they had no idea what they were supposed to do next. (Professional working for an outpatient old age care service)

One particular challenge consisted in dealing with older people suffering from dementia. Dementia sufferers being cared for by LIMCs featured prominently in our study. As described earlier, dementia is a key risk factor for abuse in care, with the risk being multiplied by a lack of knowledge about the disease and corresponding skills. This deficiency can result in questionable practices, even if these are often actually born out of a sense of concern for the people in need of care. We learned, for example, of the following practice in a household where a woman suffering from dementia was cared for in close cooperation between her husband and the LIMC. The husband reported,

And then, when she has to go to the toilet in the morning, it's, well every other day she moves her bowels, but then of course it's a palaver in the morning. She holds it in for so long; she doesn't do a number two, not in her incontinence pants. And she holds it all in, even though she really needs to go to the toilet. And she gets laxative drops and stuff, and then it's sometimes tough, but she gets sip after sip of water in the morning until she swallows it the wrong way and then she's coughing away and then it works, it loosens. I sometimes feel bad about it, but it's, there's no other way. That's how it is, but at least it works.

Outpatient old age care services can make up for the lack of professional qualifications to some extent if they are involved in the care. Professionals working for outpatient old age care services frequently reported in the interviews that the LIMCs turned to them with questions about nursing. They often also took the initiative themselves to give instructions or intervene when they found that the LIMCs were making mistakes in the care they provided. One professional working for an outpatient old age care service said,

> *And then we gave her instructions on positioning; she (the cared-for person) had bedsores in various places the whole time and then we, we had to give her instructions. So I go in and I do the nursing and go, stay out of it after the nursing and then she repositions her, she has to change her position because that costs money, of course, if we do special repositioning.*

Apart from the fact that these potential corrective measures can only take place in households where outpatient old age care services are involved, different limitations could be identified despite their involvement. A lack of time and means of communication were brought up in particular. "We [an outpatient old age care service; the authors] tried to give her [the LIMC; the authors] instructions, but she couldn't communicate with us. She used an electronic translator; we repeated things two or three times and tried to explain using gestures. But it's difficult without language" (professional working for an outpatient old age care service).

Misuse of Medication

Another potential source of abuse in care arrangements with LIMCs is the incorrect use, or the misuse, of medication. This includes incorrect dosages and the strategic use of sedatives to calm down the care recipients or send them to sleep. As they often lack training in medicine and nursing, LIMCs are actually not authorized to provide clinical care. This includes administering medication, managing wounds, measuring blood pressure, inserting and replacing nasogastric tubes and so on. However, our data indicate that family members, in fact, assign LIMCs clinical care duties. Different patterns appeared when it came to the administration of medication. In some cases, for example, family members filled the pill dispenser themselves at the beginning of the week, but delegated the administration of the medication to the LIMCs. In other cases, the LIMC was put in charge of both sorting and administering the medication.

There were also indications that family members tolerated or even encouraged questionable practices so as not to put an otherwise functioning care arrangement at risk. In the case of older care recipients with a disturbed sleep pattern or behaviors, which could quickly push the LIMC (or any other carer) to their limits, relatives looked for a solution that combined their interest in a stable care arrangement and the LIMC's need for sleep. This was seen, for example, in the use of sleeping tablets and sedatives, which was sometimes left to a certain extent in the hands of the LIMC.

However, it is not just the potentially improper use of sedatives that could cause harm. If the 24-hour caregiver deals incorrectly with the consequences of the sedation, this can also pose a risk. As one professional working for an outpatient old age care service reported,

> *The woman [the person in need of care; the authors] kept calling out a lot (…) kept calling someone's name, out of restlessness (…) inner unrest. (…) What happened in the end was, she was sedated so that the Polish woman could get some kind of sleep at night. (…) So she was prescribed Tavor and that meant she didn't move anymore and just always lay in one position, and then she developed a big sore. The woman [the LIMC; the authors] simply didn't move her.*

Alcohol Abuse

One subject which came up in the interviews with all actors involved was alcohol consumption or alcohol abuse among the LIMCs. The managing director of a placement agency expressed this as follows: "So, what often happens is that often, older people have very big stocks of wine and spirits in the cellar and, to put it very carefully, the Polish woman might be a bit susceptible to that kind of thing, so the alcohol issue comes up once and again." In view of the stress factors, which have been laid out, the regular and/or excessive consumption of alcohol can be interpreted as a coping mechanism. This was suggested in particular by the interviews with the outpatient old age care services, whose employees associated alcohol consumption with the difficult situation of the LIMCs (loneliness, being overwhelmed, etc.). However, it can also arise from the LIMCs' biographies.

One specific example in which a relative reported the consequences of alcohol abuse on the part of the LIMC is a daughter who lived around 200 km from her 90-year-old cared-for mother. She described how the LIMC was difficult to get hold of for days, as she spent most of her time in the cellar doing the washing, as she claimed. When it became impossible to contact her by telephone, the daughter asked a neighbor to check on her mother. In the daughter's words, the neighbor was confronted with the following: "She was so drunk that she couldn't do a thing anymore, she couldn't answer the phone, and my mother was still lying in her nightie with totally soaked incontinence pants, she hadn't had anything all day. It was after Christmas, she always had water by the bed and chocolate, she ate chocolates and just kept quiet. She spent all day in a dark room" (relative, 61 years old).

Abuse by the Care Recipients
Physical violence and other forms of abuse carried out by the care recipients were discussed in the interviews, albeit relatively rarely. One case, which stood out in this regard, was that of an older woman in need of care who had spoken out vehemently against entering into a care arrangement of this kind. She was of the opinion that although she might need her daughter's help, she could otherwise still live on her own. After the daughter employed an LIMC, her mother prohibited the carer from using the stove to cook, and forced her to walk at a short distance behind her when going on walks. She also deliberately got on the wrong bus to cause trouble for the LIMC. In addition, the daughter reported that her mother verbally and physically attacked the LIMC and that the frightened caregiver telephoned her for help.

Indeed, several relatives reported that the older people were resistant to the idea of an LIMC moving in, for example, because they were convinced of still being able to manage activities of daily living by themselves or because they were reluctant to the idea of having an unknown person in the same household. The following quote from an interview with an older care recipient's daughter shows the pressure which can be placed on the old people when the decision is made. This can, however, also be seen as an expression of how concerned the relatives feel or as preventing them from going into a residential care home; something which is unwanted by everyone involved. "'Mum, your only choice is between going into a nursing home or staying at home, and you always told me that you wanted to stay at home as long as possible, with your lovely balcony and that.' 'Oh, she is disturbing me' and so on, 'I don't like having someone always dancing around me' and I said 'But you know, we just can't do it anymore'" (relative, 61 years old). This example makes it clear that the older woman's right to self-determination is both being taken into account and violated: to be able to stay at home, she has to accept an unknown person in her home. Decisions against the preference of the older persons' wishes to be cared for might be a source of discontentment which by no means must result in an aggressive or violent behavior toward the LIMC but carries such a potential.

A distinction must be made between the deliberate perpetration of abuse against LIMCs and unintentional acts of abuse, for example, resulting from dementia. One family carer, for example, reported that his wife severely verbally abused a 24-hour caregiver for no reason of which he was aware. In another case, an older woman suffering from dementia was reported to have used biting to avoid going in the shower.

These examples show that abuse against LIMCs can come in many forms. It may result from people being disappointed in family members, for example, if a daughter is no longer willing or able to provide the care required, or older people may see the care arrangement as an unwanted restriction of their autonomy. The abuse of LIMCs can also be a result of mental illness. In either case, it is challenging for LIMCs to deal with these situations, which may also lead to violence and aggression. Some relatives took preventive action by avoiding certain situations which could trigger violence in people in need of care. One family, for example, decided that the LIMC only needed to give their cared-for mother a shower if she followed the request without resistance. In another case, an outpatient old age care service was paid to bathe the cared-for mother twice a week, as she had previously strongly resisted being bathed by the LIMC.

6 Discussion and Conclusion

This contribution shows that issues around abuse and protection against abuse in LIMC care arrangements must be discussed in the context of complex, interrelated factors, structures and processes. There is no clear answer as to whether these care arrangements increase or decrease the risk of abuse in care for older people. Whether risks or the potential for protection come into play depends on a combination of different factors in the care arrangement which may not only vary from one arrangement to the next but may also often differ over time or from person to person. High turnover among the LIMCs or changes in the older person's care need can also be of relevance. Based on our data, no general cause-and-effect relationships can be derived.

Nevertheless, our study indicates that abuse and protection against abuse in private households employing LIMCs are linked to and raise critical questions about a wide range of issues affecting Germany's current care system. This results, for example, from the observation that the overburdening and insufficient qualifications of those involved in caregiving can also be risk factors for violence in private households with LIMCs. It also results from the finding that the integration of these arrangements in a larger care mix may have potential for protection against violence or the alleviation of violence. Still, the financing of a larger care mix often depends on the financial resources of the individual households.

In particular, questions of abuse and protection against abuse in private households with an LIMC are connected to the widespread practice of undeclared work in this area. This is not only true of the stressors placed on the LIMC due to the precarious working conditions as factors which could potentially promote abuse. Irregular employment statuses also raise issues around whether abuse in such care arrangements is addressed and made public. The topics of abuse and irregular employment are, after all, difficult to separate in the public debate. However, so far the German state has shown little interest in focusing on the extent and the implications of irregular employment in home care for older people, instead choosing the option of "tacitly tolerating" it (Lutz & Palenga-Möllenbeck, 2010) in the knowledge that the (irregular) employment of LIMCs has become a mainstay propping up the German care system.

One expression of this policy of tolerating is that jobs in private households are hardly ever monitored (Hess, 2008; Scheiwe, 2010). By contrast, economic sectors such as the construction or catering industries, which are known for irregular employment, are constantly subject to monitoring. There are many signs of the state having an interest in keeping the lid on the practice of irregularly employing 24-hour caregivers. As the Group of Experts on "World Economy and Social Ethics" (2015: 25) argues,

> *By tacitly condoning the phenomenon and failing to apply any regular, comprehensive monitoring, governments tolerate these mostly irregular employment relationships which are far from offering the level of protection provided under national labor law. They thus see themselves as absolved from any need to develop solutions for the deficit of care workers. As such, the employment of domestic workers in many Western societies shores up a form of organizing the care work that is required which at first glance appears very inexpensive for the state. (Translation by the authors)*

Added to this is the fact that abuse and protection against abuse in care arrangements with LIMCs must generally be viewed against the backdrop of the widespread taboo surrounding domestic abuse against older people in Germany. Among other things, this is reflected in a lack of any social infrastructure which could address, make public and process the subject of abuse against older people (Schweppe, 2018). Today, various social institutions can be found, some embodied in the law, which deal with domestic abuse against young people (children and adolescents) and women. They are aimed both at the victims and at people who have uncovered or suspect domestic abuse. Such institutions and initiatives are largely lacking for older people. The police are usually the only organization people can turn to. However, the police are prepared to deal with questions of elder abuse and are very limited in their possibilities of intervention in case of elder abuse. In addition, under the conditions of irregular employment relationships, the police are hardly a suitable option either for the relatives or for the LIMCs: in both cases, reporting an issue or seeking help from the police would likely bring the irregular employment into the open.

The first step worth recommending would appear to be to legalize the employment relationships. However, without any other fundamental changes to the way care for older people is financed, legalization would by no means be an appropriate solution from the point of view of the private households. It would significantly increase the cost of these care arrangements, preventing a considerable number of households from making use of this option. Private households, which chose to employ an LIMC irregularly to save money, would also run the risk of being criminalized. The question of whether these employment relationships could be legalized is thus closely linked to whether they can be afforded, regulated and monitored. The problems of how to distinguish between and document working hours and time off in such care arrangements remain unsolved, for example. The same applies to the issue of the LIMCs' qualifications, for which there is yet no binding standard.

Initiatives, such as those by individual placement agencies or welfare associations, offer concepts in these contexts that could help protect against elder abuse in care arrangements involving LIMCs. Examples of such initiatives include the CariFair (Caritas) and FairCare (Diakonie) programs. These programs help with finding and placing LIMCs at the level of various district associations. The prerequisite is that the families employ the LIMC regularly and thus pay taxes and social security contributions. They are also required to pay tariff salary, comply with the law on working hours and involve an outpatient old age care service (Caritas) in their arrangements. Currently, 500–600 LIMCs have been placed through CariFair (see Menebröcker in this book). This relatively low number can be explained by the comparatively high costs incurred by the model of employment and the additional services required.

Regular employment conditions, regulated working hours and training certificates raise the costs of (private) care. As a result, better protection would only be accessible to private households which have the necessary financial means. This would thus further exacerbate the social inequalities which already characterize the care system in Germany. Even as a "model for the few," it would only be of limited interest to those market operators which focus on growth and profit, such as the placement agencies. It thus comes as no surprise that a sector-based group of placement agencies has been working for some time to broaden the financing basis for care arrangements involving LIMCs. However, due to their employment models, the placement agencies themselves are repeatedly criticized for operating in a legal gray zone (see Emunds, 2016). To assert their interests, they thus initially depend on the state creating legal certainty with regard to their actions.

Note

1. The research project was carried out jointly by Johannes Gutenberg University Mainz, Germany (Vincent Horn, Cornelia Schweppe), and Radboud University, Netherlands (Anita Böcker, María Bruquetas-Callejo), with funding from the German Research Foundation (DFG) and the Dutch Research Council (NWO) as part of the Open Research Area (ORA) program (2016–2019).

References

Apitzsch, U., & Schmidbauer, M. (2010). *Care und Migration: Die Ent-Sorgung menschlicher Reproduktionsarbeit entlang von Geschlechter- und Armutsgrenzen: Die Ent-Sorgung menschlicher Reproduktionsarbeit in der globalen Peripherie*. Verlag Barbara Budrich.

Arend, S., & Klie, T. (2017). *Wer pflegt Deutschland. Transnationale Pflegekräfte—Analysen, Erfahrungen, Konzept*. Vincentz Verlag.

Ayalon, L. (2009). Fears come true: The experiences of older care recipients and their family members of live-in foreign home care workers. *International Psychogeriatrics, 21*(4), 779–786.

Ayalon, L. (2011). Abuse is in the eyes of the beholder: Using multiple perspectives to evaluate elder mistreatment under round-the-clock foreign home carers in Israel. *Ageing & Society, 31*, 499–520.

Bestmann, B., Wüstholz, E., & Verheyen, F. (2014). *Pflegen: Belastung und sozialer Zusammenhalt. Eine Befragung zur Situation von pflegenden Angehörigen*. Techniker Krankenkasse.

Böcker, A., Horn, V., & Schweppe, C. (2017). Old age care regimes and the emergence of transnational long-term care arrangements for the elderly. In L. Good Gingrich & S. Köngeter (Eds.), *Transnational social policy, social welfare in a world on the move* (pp. 222–242). Routledge.

Böning, M., & Steffen, M. (2014). *Migrantinnen aus Osteuropa in Privathaushalten*. ver.di.
Buchegger-Traxler, A. (2016, September). Risiko Abhängigkeit. Gewalt und Vernachlässigung im privaten Nahbereich älterer Menschen. *Informationsdienst des österreichischen Instituts für Familienforschung*, 1–4.
Cooney, C., Howard, R., & Lawlor, B. (2006). Abuse of vulnerable people with dementia by their carers: Can we identify those most at risk? *International Journal of Geriatric Psychiatry, 21*, 564–571.
Cooper, C., Selwood, A., Blanchard, M., Walker, Z., Blizard, R., & Livingston, G. (2009). Abuse of people with dementia by family carers: Representative cross sectional survey. *BMJ, 338*, b155. https://doi.org/10.1136/bmj.b155
Dieck, M. (1987). Gewalt gegen ältere Menschen im familiären Kontext—Ein Thema der Forschung, der Praxis und der öffentlichen Information. *Zeitschrift für Gerontologie, 20*, 305–313.
Eggert, S., Schnapp, P., & Sulmann, D. (2018). Aggression und Gewalt in der informellen Pflege. Retrieved January 2020, from https://www.zqp.de/wpcontent/uploads/ZQP_Analyse_Gewalt_informelle_Pflege.pdf
Emunds, B. (2016). Menschenunwürdige Pflegearbeit in deutschen Privathaushalten. Sozialethische Bemerkungen zu den Arbeitsverhältnissen mittel- und osteuropäischer Live-Ins. *Jahrbuch für Christliche Sozialwissenschaften, 57*, 199–224.
Görgen, T. (2017). Wissen über das Phänomen Gewalt in der Pflege. In *ZQP Report Gewaltprävention in der Pflege* (pp. 8–12). Zentrum für Qualität in der Pflege.
Görgen, T., Herbst, S., Kotlenga, S., Nägele, B., & Rabold, S. (2012). Kriminalitäts- und Gewalterfahrungen im Leben älterer Menschen: Zusammenfassung wesentlicher Ergebnisse einer Studie zu Gefährdungen älterer und pflegebedürftiger Menschen. Bundesministerium für Familie, Senioren, Frauen und Jugend (BMFSFJ). Berlin.
Graß, H., Walentich, G., Rothschild, M. A., & Ritz-Timme, S. (2007). Gewalt gegen alte Menschen in Pflegesituationen. Phänomenologie, Epidemiologie und Präventionsstrategien. *Rechtsmedizin, 17*(6), 367–371.
Hess, S. (2008). Au Pairs als informalisierte Hausarbeiterinnen. In C. Gather, B. Geissler, & M. S. Rerrich (Eds.), *Weltmarkt Privathaushalt. Bezahlte Hausarbeit im globalen Wandel* (2nd ed., pp. 103–119). Dampfboot Verlag.
Hielscher, V., Kirchen-Peters, S., & Nock, L. (2017). *Pflege in den eigenen vier Wänden: Zeitaufwand und Kosten. Pflegebedürftige und ihre Angehörigen geben Auskunft*. Hans-Böckler-Stiftung.

Hirsch, R. D. (2011). Konflikte in Pflegebeziehungen: Eine Herausforderung für Pflegende und die Gesellschaft. In T. Schürmann, M. Geuther, & L. Thaut (Eds.), *Alt und Jung: vom Älterwerden in Geschichte und Zukunft* (pp. 137–151). Förderverein des Freilichtmuseums am Kiekeberg.

Horn, V., Schweppe, C., Böcker, A., & Bruquetas-Callejo, M. (2019). Live-in migrant care worker arrangements in Germany and the Netherlands: Motivations and justifications in family decision-making. *International Journal of Ageing and Later Life, 13*(2), 83–114.

Ignatzi, H. (2014). *Häusliche Altenpflege zwischen Legalität und Illegalität: dargestellt am Beispiel polnischer Arbeitskräfte in deutschen Privathaushalten.* LIT Verlag.

Karakayali, J. (2010). *Transnational Haushalten. Biografische Interviews mit „care workers" aus Osteuropa.* VS Verlag für Sozialwissenschaften.

Kniejska, P. (2016). *Migrant care workers aus Polen in der häuslichen Pflege: Zwischen familiärer Nähe und beruflicher Distanz.* Springer VS.

Konopik, N., Halscheidt, A., Schulz, L., & Oswald, F. (2018). *Interdisziplinäre Untersuchung zu Rechtsschutzdefiziten und Rechtsschutzpotentialen bei Versorgungsmängeln in der häuslichen Pflege alter Menschen (VERA). Sozialwissenschaftliche Analyse zum Projekt.* Retrieved September 2019, from https://www.pflegebevollmaechtigter.de/files/upload/pdfs_Veranstaltungen/04_VERA-Sozwiss.%20Analyse-2019-04-10-send.pdf

Krawietz, J. (2014). *Pflege grenzüberschreitend organisieren. Eine Studie zur transnationalen Vermittlung von Care-Arbeit.* Frankfurt am Main: Mabuse-Verlag GmbH.

Kühnert, S. (1997). Ursachen und Bedingungsfaktoren für Gewalt. In *Gewalt gegen Ältere.* Dokumentation einer Fachtagung. BMFSFJ.

Lachs, M. S., & Pillemer, K. A. (2015). Elder abuse. *New England Journal of Medicine, 373*, 1947–1956.

Lutz, H., & Palenga-Möllenbeck, E. (2010). Care work migration in Germany: Semi-compliance and complicity. *Social Policy & Society, 9*(3), 419–430.

Maibaum, M. (2017). Hilfe bei der Pflege daheim: "Dann holt Euch doch 'ne Polin". NRZ-Pflegeserie. Retrieved April 16, 2019, from https://www.nrz.de/region/dann-holt-euch-doch-ne-polin-id210073997.html

Metz-Göckel, S., Koch, A., & Münst, A. S. (2006). Leben in zwei Welten: zur Pendelmigration polnischer Frauen. *Zeitschrift für Frauenforschung & Geschlechterstudien, 24*, 51–68.

Müller-Hergl, C. (2011). Gewalt und Vernachlässigung (Elder Abuse): Stand der Diskussion. Arbeitspapier DZD 009. Dialog- und Transferzentrum Demenz.

Nägele, B., Kotlenga, S., Görgen, T. & Mauder, B. (2010): Ambivalente Nähe: eine qualitative Interviewstudie zur Viktimisierung Pflegebedürftiger in häuslichen Pflegearrangements. In: Thomas Görgen (Hrsg.): Sicherer Hafen oder gefahrvolle Zone? Kriminalitäts- und Gewalterfahrungen im Leben alter Menschen. (S. 208-480). Frankfurt a.M.: Verlag für Polizeiwissenschaft.

Neuhaus, A., Isfort, M., & Weidner, F. (2009). *Situation und Bedarfe von Familien mit mittel- und osteuropäischen Haushaltshilfen*. Deutsches Institut für angewandte Pflegeforschung e.V.

Orfila, F., Coma-Sole, M., Cabanas, M., Cegri-Lombardo, F., Moleras-Serra, A., & Pujol-Ribera, E. (2018). Family caregiver mistreatment of the elderly: Prevalence of risk and associated factors. *BMC Public Healthcare, 18*(167), 1–14.

Petermann, A., Ebbing, T., & Paul, M. (2017). *Das Tätigkeitsprofil von Betreuungspersonen in häuslicher Gemeinschaft*. Berufsakademie für Gesundheits- und Sozialwesen Saarland.

Pillemer, K., Burnes, D., Riffin, C., & Lachs, M. (2016). Elder abuse: Global situation, risk factors, and prevention strategies. *The Gerontologist, 56*(2), 194–205.

Pillemer, K., & Suitor, J. (1992). Violence and violent feelings: What causes them among family caregivers? *Journal of Gerontology: Social Sciences, 47*(4), 165–172.

Pinquart, M. (2016). Belastungs- und Entlastungsfaktoren pflegender Angehöriger—die Bedeutung der Erwerbstätigkeit. In *Vereinbarkeit von Beruf und Pflege* (pp. 60–72). Zentrum für Qualität in der Pflege.

Rabold, S., & Görgen, T. (2007). Misshandlung und Vernachlässigung älterer Menschen durch ambulante Pflegekräfte: Ergebnisse einer Befragung von Mitarbeiterinnen und Mitarbeitern ambulanter Dienste. *Zeitschrift für Gerontologie und Geriatrie, 40*, 366–374.

Rossow, V., & Leiber, S. (2017). Zwischen Vermarktlichung und Europäisierung: Die wachsende Bedeutung transnational agierender Vermittlungsagenturen in der häuslichen Pflege in Deutschland. *Sozialer Fortschritt, 66*, 285–302.

Satola, A. (2015). *Migration und irreguläre Pflegearbeit in Deutschland.* Fachhochschule Fulda.

Scheiwe, K. (2010). Die soziale Absicherung häuslicher Pflege über Grenzen hinweg—Rechtliche Grauzonen, (Ir-)Regularität und Legitimität. In K. Scheiwe & J. Krawietz (Eds.), *Transnationale Sorgearbeit: Rechtliche Rahmenbedingungen und gesellschaftliche Praxis* (pp. 123–150). VS Verlag.

Scheiwe, K., & Krawietz, J. (2010). *Transnationale Sorgearbeit. Rechtliche Rahmenbedingungen und gesellschaftliche Praxis*. VS Verlag.

Schweppe, C. (2018). Öffentliches Bewusstsein für Gewalt in der häuslichen Pflege. In BAGSO: Dokumentation: Gewaltprävention in der häuslichen Pflege. Expert workshop by BAGSO. Retrieved July 15, 2020, from https://www.bagso.de/fileadmin/user_upload/bagso/03_Themen/Pflege/Angehoerige/doku_expertenworkshop_gewaltpraevention_pflege.pdf

Statistisches Bundesamt. (2017). *Pflegestatistik 2015. Pflege im Rahmen der Pflegeversicherung. Deutschlandergebnisse.*

Stiftung Warentest. (2017). Pflege zu Hause. Die besten Vermittler für Betreuungskräfte aus Osteuropa. *Test, 5*(17), 86–95.

Tießler-Marenda, E. (2012). Rahmenbedingungen für die Arbeit in Pflegehaushalten in Deutschland. In A. Hitzemann, N. Schirilla, & A. Waldhausen (Eds.), *Pflege und Migration in Europa. Transnationale Perspektiven aus der Praxis* (pp. 103–115). Lambertus.

Wang, M., Sun, H., Zhang, J., & Ruan, J. (2018). Prevalence and associated factors of elder abuse in family caregivers of older people with dementia in central China cross-sectional study. *International Journal of Geriatric Psychiatry, 34*, 299–307.

Weissenberger-Leduc, M., & Weiberg, A. (2011). *Gewalt und Demenz, Ursachen und Lösungsansätze für ein Tabuthema in der Pflege*. Springer Verlag.

Wiglesworth, A., Mosqueda, L., Mulnard, R., et al. (2010). Screening for abuse and neglect of people with dementia. *Journal of the American Geriatrics Society, 58*, 493–500.

World Health Organization (WHO). (2002). *World report on violence and health*. WHO.

Index[1]

[1] Note: Page numbers followed by 'n' refer to notes.

V. Horn et al. (eds.), *The Global Old Age Care Industry*,
https://doi.org/10.1007/978-981-16-2237-3

The manufacturer's authorised representative in the EU is Springer Nature Customer Service Centre GmbH, Europaplatz 3, 69115 Heidelberg, Germany. If you have any concerns regarding our products, please contact ProductSafety@springernature.com

Printed and bound by CPI Group (UK) Ltd, Croydon, CR0 4YY

30/09/2026

02225326-0016